LAPSES IN MATHEMATICAL REASONING

V. M. BRADIS, V. L. MINKOVSKII
and A. K. KHARCHEVA

Translated by
J. J. SCHORR-KON

English Translation Edited by
E. A. MAXWELL

DOVER PUBLICATIONS, INC.
Mineola, New York

Published in Canada by General Publishing Company, Ltd., 30 Lesmill Road, Don Mills, Toronto, Ontario.

Bibliographical Note

This Dover edition, first published in 1999, is an unabridged and unaltered republication of the English translation of the second Russian edition (1959), as published in "The Commonwealth and International Library of Science, Technology, Engineering and Liberal Studies" by Pergamon Press, Oxford and London, and the Macmillan Company, New York, in 1963.

Library of Congress Cataloging-in-Publication Data

Bradis, V. M. (Vladimir Modestovich), 1890–
[Oshibki v matematicheskikh rassuzhdeniiakh. English]
Lapses in mathematical reasoning / V.M. Bradis, V.L. Minkovskii, and A.K. Kharcheva; translated by J.J. Schorr-Kon ; English translation edited by E.A. Maxwell.—Dover ed.
p. cm.
Originally published: Oxford : Pergamon Press ; New York : Macmillan, 1963.
Includes bibliographical references.
ISBN 0-486-40918-X (pbk.)
1. Mathematics—Problems, exercises, etc. 2. Problem solving. I. Minkovskii, V.L. (Vladimir L'vovich) II. Kharcheva, A.K. (Avgusta Konstaninovna) III. Title.

QA43 .B6953 1999
510'.76—dc21

99-045828

Manufactured in the United States of America
Dover Publications, Inc., 31 East 2nd Street, Mineola, N.Y. 11501

Contents

From the Preface to the First Edition

INFINITELY varied lapses occur and keep occurring in the course of mathematical arguments. It is worthwhile to discuss with high-school students some of these errors for two reasons: first, by thoroughly familiarizing ourselves with a mistake we protect ourselves from repeating it in the future; second, it is easy to make the very process of looking for the mistake fascinating for the student, and the study of lapses becomes a means of stimulating interest in the study of mathematics.

In the majority of cases it is easy to lead an argument, in which a given lapse is introduced towards a clearly false conclusion. This gives the semblance of a proof of some obvious absurdity, or a so-called *sophism*. To analyse a sophism means to point to the lapse which has been introduced in the argument and because of which the absurd conclusion has been made.

Many such erroneous arguments from various branches of mathematics are known and there exist several compilations of them. The present collection is meant for high-school students and contains material of varying difficulty which may be proposed by teachers at most levels. It is valuable to make use of this material in the activities of school mathematics clubs, but some of the problems may be profitably analysed also in the course of ordinary classes, particularly while revising.

We note that in the course of analysing lapses it is absolutely necessary to press for complete clarity: the students should establish clearly wherein consists the lapse contained in the argument and how it is to be corrected. Taking this into account, the authors have supplied a detailed explanation after each erroneous argument cited in the present collection. Of course, this explanation should not be read immediately after studying the problem, but after persistent attempts to gain an understanding of it independently. In many cases the reader will find the explanation independently, or after a few hints from the teacher. Particular attention should be given to accurate formulation. The point is, that insufficient accuracy in the verbal formulation of a theorem, common among students, may sometimes be the basis of a misunderstanding. (A good example of such a misunderstanding is given in § 1 of chapter III*.) Such inaccuracies are met not only in the answers of students but also in commonly accepted formulations. . . .

The work of A. K. Kharcheva "Mathematical Sophisms and Their Application in the School," presented by her as her thesis† for the diploma at the Kalinin Pedagogical Institute, forms the basis of the present collection. The final editing of the book and some additions to the original text are by V. M. Bradis.

1937

V. M. Bradis
A. K. Kharcheva

* In the present edition this forms section 7 of the second part of Chapter I.

† At that time, before the introduction of state examinations, the graduates of pedagogical institutes had to produce and defend a diploma thesis.

Preface to the Second Edition

THE book by V. M. Bradis and A. K. Kharcheva *Lapses in Mathematical Reasoning*, published in 1938 and long out of print, enjoyed in its time, a considerable success among teachers. According to an agreement with the authors I have undertaken its revision for publication. In preparing the new edition I have made use of my paper "An Attempt at Classification of Exercises on Refutation of False Mathematical Arguments," printed in 1956 in the *Academic Records of the Chairs of the Faculty of Physics and Mathematics of the Orlov State Pedagogical Institute* (vol. XI, no. II, pp. 122–148). Also some of the less successful chapters of the first edition of the book are omitted, a few new erroneous arguments are added and the explanations are carried out in separate parts of the respective chapters.

In the present book the false proofs are distributed according to a scheme of classification by subject-matter. This means, that the traditional division of the material into arithmetic, algebra, geometry and trigonometry is retained, but within parts of the school mathematical curriculum, the division of the examples is carried out in accordance with the classification set forth in the first chapter.

In putting together the present compilation the authors have made use of various manuals, among them:

Obreymov, V. I., *Mathematical Sophisms* (Matematicheskiye sofizmy), 3rd ed. Petersburg (1898).

Goryachev, D. N., and A. M. Voronetz, *Problems, Questions and Sophisms for Mathematics Lovers* (Zadachi, voprosy i sofizmy dlya lyubitelei matematiki). Moscow (1903).

Lyamin, A. A., *Mathematical Paradoxes and Interesting*

Problems (Matematicheskiye paradoksy i interesnyye zadachi). Moscow (1911).

Lyanchenkov, M. S., *Mathematical Anthology* (Matematicheskaya Khrestomatiya). Petersburg (1911–1912).

I hope that those who have familiarized themselves with the book and have comments thereon will direct them to the mathematical editors of the Uchpedgiz (State Publishing House for Teaching and Pedagogical Literature) at the following address: Moscow, 1–18, 3–ii proezd Mar'inoi roshchi, dom 41.

V. L. Minkovskii

CHAPTER I

Exercises in Refuting Erroneous Mathematical Arguments and their Classification

Introduction

In science every positive or negative assertion may be called a *thesis*. For example, in proving some theorem, we have a thesis—the text of that theorem.

To prove a thesis means to establish its truth. To refute a thesis means to demonstrate its falsity.

The verification of a thesis consists in its proof or refutation.

The refutation of a proof does not necessarily imply the refutation of the thesis. If the thesis is true, then the refutation of the proof testifies only to the fact that incorrect arguments are brought in its defence or else that an error in the argument is introduced. However, the truth of the thesis remains in question as long as the necessary arguments are not presented together with a logically faultless statement of proof.

In checking a proof supporting a true or apparently true thesis, it is by no means easy always to note the presence of an error. The problem becomes much easier if, knowing that an error is actually present in the proof, we start with the particular aim of exhibiting it.

If the thesis expresses a false opinion, then any proof of that thesis will always be false. The ability to refute the proof of a thesis in the case of falsity is just as necessary as the ability to prove a thesis in the case of accuracy.

In the course of political, scientific and everyday disputes, in the process of a court investigation and analysis, in attempts to solve various problems, one must learn not only to prove, but also to refute.

V. I. Lenin, analysing conscious and subconscious errors in the domain of logical thought of his political adversaries, used to recall those arguments "which are called mathematical sophisms by mathematicians and in which—in a way that, on the face of it, is strictly logical—it is proven that twice two makes five, that a part is greater than the whole, etc."; and he used to point out that "there exist collections of such mathematical sophisms, and they bring profit to schoolchildren."*

In the methodological circular of the Ministry of Education RSFSR "On the Teaching of Mathematics in the grades V–X" (1952, p. 41) it is indicated that "a very useful help for the development of the logical abilities of pupils is given by all kinds of sophisms."

I. Mathematical Sophisms and their Pedagogical Value

Sophism is a word of Greek origin, meaning in translation a wily fabrication, trick, or puzzle. The matter concerns a "proof," aimed at a formally logical establishment of an absurd premise.

Mathematical sophisms involve lapses in mathematical arguments where, even though the result is patently false, the errors leading to them are more or less well concealed. To uncover a sophism means to show the error in the argument by means of which the outer appearance of proof has been created. The demonstration of the error is usually arrived at by counterposing the correct argument against the false one.

In the main, mathematical sophisms are constructed on the basis of incorrect usage of words, on the inaccuracy of formulations, very often on neglecting the conditions of applicability of

* V. I. Lenin, *Works*, Vol. 7, 4th Ed., p. 78. Moscow, 1946.

theorems, on a hidden execution of impossible operations, on invalid generalizations, particularly in passing from a finite number of objects to an infinite one, and on masking of erroneous arguments or assumptions by means of geometrical "obviousness". V. I. Lenin gives a general formulation characterizing sophistry as ". . . the grasping of the outer coincidence of cases outside the relation of events. . . ."*

The intricacy of a mathematical sophism increases with the subtlety of the error concealed in it, with the lack of advance warning in ordinary school instruction on the subject, and with the artfulness of its concealment by inaccuracies of verbal expression. For purposes of concealment one usually complicates the plot of the sophism, i.e. one formulates a situation, for the proving of which one must use some true mathematical propositions, bringing about a distraction which makes the reader look for the error along a false path. In some sophisms this kind of distraction is successfully aided by an optical illusion.

The basic aim of introducing sophisms in school, lies in accustoming the students to critical thinking, to knowing not only how to carry out definite logical schemes and definite processes of thought, but also how to examine critically every stage of an argument in accordance with established principles of mathematical thought and computational practice.

In the opinion of experienced teachers the possibilities of effective applications of mathematical sophisms increase as the students advance through school and their interest in the logical structure of science grows. This work may be proposed in a particularly profound and useful form before a mathematical club of students of the higher grades, where a heightened interest in the logical bases of the methods of logical proof usually shows itself.

Mathematical sophisms require that their texts be read with particular attention and great caution. In studying them one has to seek carefully the proper accuracy of statement and notation, the observance of all the conditions governing theorems, the

* V. I. Lenin, *Works*, Vol. **21**, 4th Ed., p. 100. Moscow, 1948.

absence of inadmissible generalizations, forbidden operations, reliance on apparent properties of figures in auxiliary constructions. All these points are methodologically valuable, since they are aimed at a complete mastery of the subject in contradistinction to a formal one, which "is characterized by undue domination in the consciousness and memory of the students of the accustomed outer (verbal, symbolic, or pictorial) expression of the mathematical fact over the content of that fact" (A. Ya. Khinchin).

The degree of mastery of a mathematical fact is considerably strengthened when its perception is stimulated by the absurd allegation contained in the formulation of the sophism.

Exercises in uncovering sophisms do not guarantee the absence of similar errors in the students' own arguments, but they do give the possibility of uncovering and understanding more quickly any error that may appear. This thought, as applied to pedagogy, consists in the fact that mathematical sophisms proposed to students should, as a rule, be used not only to prevent errors, but to check the degree of familiarity and firmness of grasp of the given material. Upon this proposition is based the working practice of our better teachers of mathematics who, to some extent, use sophisms in the concluding stages of Exercises to a given Chapter and in revision.

The pedagogue prevents students' mistakes by a thorough analysis of the concepts studied in class. The teachers' own familiarity with typical students' errors, their origin, and the material of mathematical sophisms helps towards a better attainment of this goal. The degree of the teacher's preparation in this direction is usually shown in the choice of examples and in the clarification of existing variations of a given type, with the aim of preventing the appearance of one-sided associations and incorrect generalizations.

Most teachers agree that in explaining new material, one should, as a rule, avoid fixing the attention of the students upon errors about to appear, in order not to create false intuitive impressions.

Pedagogically warranted use of mathematical sophisms does

not exclude the formulation of problems in misleading form, but, on the contrary, often uses it as a preliminary stage of the work as a source of instructive errors. For these problems the student finds no ready-made answers in the teacher's textbook. What is here required of the student is an understanding of the essence of the theoretical material studied, independent thought, and deliberate operation of the known stock of mathematical facts. Some such problems are:

1. When is $\frac{a}{b}$ equal to unity?
2. From the fact that $a > b$, is it possible to conclude that $|a| > |b|$?
3. From the equality $(a - b)^2 = (m - n)^2$, may one draw the conclusion that $a - b = m - n$?
4. Does the formula
$$\sqrt{(x)} \times \sqrt{(y)} = \sqrt{(xy)}$$
hold for all values of x and y?
5. Is the identity
$$\log x^2 = 2 \log x$$
valid for all positive values of x?
6. Define the meaning of the symbol $\vee$ in the notation $2a \vee a$, putting a equal to: (a) $\log \frac{1}{2}$, (b) $\log \cos \alpha$, where $0 \leqslant \alpha < 90°$, and generalize.
7. For what values of x do the following expressions lose their meaning?
$$\frac{x^3 - 1}{x - 1}, \quad \frac{1}{x^2 - 1}, \quad \frac{3}{\cos x}, \quad \frac{x}{\log x}.$$
8. Establish the error and correct the statement of a theorem given by a student of geometry: "A straight line parallel to one of the sides of a triangle cuts from it a triangle similar to the given one."
9. In a right-angled triangle can a median dropped to an arm coincide with the bisectrix?

It is necessary, when applying any mathematical sophism, to instruct a pupil that there should be at his disposal the requisites for uncovering that sophism. The nonobservance of this necessary condition not only completely invalidates the use of sophisms, but also makes them harmful: the student, not being able to find his way in the essence of the problem, helplessly catches hold of external methods, reducing his work to simple guesswork, loses his equilibrium, and develops streaks of indecision. All this, of course, has nothing in common with the problem of gradual and persistent testing of caution in assertions, or with the acknowledged necessity to understand the conditions of a problem and the means for its effective solution. At every point of the course, too, the teacher should be completely candid with the pupil, openly pointing out to him those logical gaps in his exposition which are the result of deliberate pedagogical adjustment.

II. Classification of Exercises in Refuting False Mathematical Arguments

In the history of the development of science an essential role was played by mathematical sophisms (once called paradoxes). They demanded increased attention to the requirements of pithy analysis and of strict proof and have led early to a prolonged prohibition (at least, official) of the use of those concepts and methods which were still not accessible to strict logical elaboration. This makes it easy to understand the early interest in the study, systematization, and pedagogical application of patently false proofs.

The recognition of the pedagogical role of mathematical exercises refuting false proofs suggests an effort to find and characterize their basic forms as a necessary condition if there is to be a rational choice and application of this material in school.

The first attempt to set up a compilation of geometrical sophisms had been understood by the author of the *Principles*, Euclid of Alexandria. Regrettably, this work of Euclid's, bearing the name of *Pseudaria*, is considered as irretrievably lost. Proclus

(410–485) tells us of its purpose and contents. From his words it is apparent that the work was meant for beginners in geometry. Its aim was to teach students to recognize false conclusions and thus be able to avoid them.

For recognizing errors Euclid sets forth ingenious methods, which he enumerates in a definite order, accompanying them by corresponding exercises. To a false proof Euclid contrasts the correct one and shows that sometimes intuition may serve as the source of error.

The eminent Russian mathematician and pedagogue V. I. Obreimov (1843–1910) has proposed his "attempt at grouping" exercises of such types and has enumerated these groups.

The first three groups of Obreimov's classification (equality of unequals, inequality of equals, and smaller exceeding the greater) embrace those false proofs whose theses contradict the application of the criteria of comparison of magnitudes, i.e. of the concepts greater than, less than, and equal to.

A fourth group consists in geometric absurdities. In it are included deductions in which an absurd conclusion follows from an error in the diagram in the presence of an irreproachable execution of all remaining logical arguments.

A fifth group is formed by "imaginary is real." Here are grouped false proofs connected with the incorrect treatment of the concept of complex numbers.

Obreimov's classification is not free of imperfections. First, in enumerating the forms of the concept being classified, the principle of uniqueness of the basis for classification of the errors is not sustained. Instead there is, on the one hand, the criteria of comparison, and, on the other, the inclusion in geometry or in some portion of the algebra course (complex numbers). Secondly, as the basis of the first three groups of Obreimov's classification, a purely external, rather general criterion of classification has been chosen, which is quite inessential as far as the characterization of false proofs is concerned. Because of this, material referring to the clarification of a given error is distributed over different divisions. Misunderstandings in connexion with division by zero are expounded in the first and second divisions, whereas misunderstandings connected with failing to change the sign in multiplying both parts of an inequality by a negative number are set forth in the second and third chapters, and so on.

A positive feature of Obreimov's classification is the separation into a distinct group of false proofs based on errors in construction.

We now proceed to study the proposals for a classification of false proofs by the German scientist Herman Schubert (1848–1911). He calls for a distinction of four forms of false proofs, based on division by zero, on the ambiguity of the square root, on geometrical deception (error in construction), and on ascribing an infinitely great value to the sum of an infinite set of numbers.

What is important in Schubert's classification is the division of false proofs according to the character of the errors which lead to them. However, classification according to the principles chosen has remained undeveloped. The four forms of false proofs enumerated by Schubert do not exhaust even the minimal content of the concept under consideration. In particular, nothing was said about false proofs constructed on the basis of a mistaken trust in geometrical intuition in those cases when there is no direct geometrical deception.

The French pedagogue and historian of mathematics E. Fourier relegates to geometrical sophisms all those sophisms whose formulations relate to geometrical objects. Thus, he includes here also geometrical embodiments of algebraic sophisms, based on the presence of purely algebraic errors. Of course, these errors may be masked not only by various geometrical assumptions but also by assumptions referring to other branches of mathematics. It is clear that a classification based on the outward form of the erroneous argument is a purely external classification.

The geometrical sophisms in the broad sense indicated are divided by Fourier into two forms: those based on errors in construction and those based on errors in argument. In the errors of argument Fourier distinguishes between errors connected with deviation from precise definitions and those connected with the carrying out of inadmissible operations on numbers.

We now give our attempt to classify exercises about refuting false mathematical arguments into the more important classes of errors in speech and thought which, usually without even naming them, the teacher stubbornly fights in his daily work both in the teaching of theory and in carrying out the exercises.

The proposed classification is aimed, to begin with, at the teacher. With this goal in view, the very name emphasizes the specific pedagogical intention of a given exercise for each form of error, thus creating the possibility of fast orientation in the material and preventing its unsystematic use.

Our teaching experience allows us to assert that this classification is useful also to students of the higher grades, whose arguments begin to be executed not only in correspondence with definite principles but also on the basis of consciousness of these principles. During this transition period the students feel a continual need to verify the accuracy of their knowledge, their logical development, the degree of their comprehension, and the adequacy of their verbal expression.

We are far from the thought that the proposed classification is free from imperfections which are characteristic, for example, of the classification of arithmetical problems according to the essential peculiarities of the methods of their solution. The use of a wider working experience and criticism will help to suggest the necessary corrections. However we take the liberty of thinking that, even in its present form, it represents a certain step forward in comparison with existing classifications pursuing, just as this one, purely pedagogical aims.

We now go on to the actual consideration of the problem.

1. Incorrectness of speech

A systematic analysis of sophisms was first given by Aristotle (384–322 B.C.) in a special treatise devoted to the refutation of sophistries in which all the errors are divided into two classes: "incorrectnesses of speech" and errors "outside speech," i.e. in thinking.

There is no need to prove that every lesson correctly planned and executed in the subject of mathematics is at the same time also a lesson in developing the powers of exposition of the students. On the pages of methodological literature one can often see the emphasis of the favourable influence of mathematics upon the improvement of the student's gift of speech, in the sense of its precision and consistency. However, these aims are not attained automatically. To attain them, daily work is necessary on the part of the mathematics teacher as he watches the student's choice of words, the form of expression of his thoughts, both in verbal answers and in written tasks. The intensive elimination of faults encountered in the students' speech involves interesting the students themselves in correcting the answers of their fellows. It should be distinctly brought to the attention of the students that irregularities of speech not only make the study of mathematics more difficult but are also responsible for numerous other fallacies.

Ambiguities of a word

As a rule, every concept in mathematics is denoted by its separate term. In exceptional cases, however, i.e. when one and the same term is used in different senses, special stipulations are necessary as to the sense in which a given term is used, if it is not clear from the context itself. To the number of ambiguous mathematical terms belong, for example, the following: square (the exponent of a power and the geometrical figure), root (in the sense of a solution of an equation and as the synonym of the word radical), number (cardinal and ordinal, abstract and concrete, exact and approximate).

EXAMPLE. Father and son arrived in a town to live there permanently. The boy, knowing from his parent's stories that there were 25,000 inhabitants in the town, hastened to announce at the station that now there were 25,002. The father laughed and began to explain something to the son. What did the father say?

Ambiguity of pronunciation

Here the matter concerns the distortion of the original sense of the sentence because of a misplaced stress in some word.

EXAMPLE. The Russian for crow (a bird) in the genitive case plural is *sorok*. The same word also means forty. Hence, the ambiguous construction "100 crows + 100 crows = 200 crows" can also mean "140 + 140 = 280."

Ambiguity of construction

Here the proposition is constructed as to admit different interpretations of its meaning.

EXAMPLE. How much is three times three and seven?

This sentence allows two different mutually exclusive orders of operations, namely: $3 \times 3 + 7$ and $3 \times (3 + 7)$.

Error of distribution

This error holds when to a term applied in a collective sense is given the meaning of a disjunctive one.

EXAMPLE. All the angles of a triangle are equal to two right angles.

Here the word "all" is applied in the sense of "sum." However the choice of the term is unfortunate since it may be understood also in the sense "every." The thought becomes absurd: "Every angle of a triangle is equal to two right angles."

Error of composition

This error is opposite to the preceding one. It arises when to a term used in a disjunctive sense is given a collective meaning.

EXAMPLE. All the angles of a triangle are less than two right angles. Here the word "all" is used in the sense "each." However the choice of the term may not be considered as appropriate, since it may be understood also in the sense "sum." In a system of Euclidean geometry the thought becomes absurd: "The sum of the angles of a triangle is less than two right angles."

We begin the analysis of errors "outside of speech" with false proofs constructed on the basis of hurried unexamined generalizations. The consideration of some of them in school is very useful, since "first of all there is a definite tendency to the extended interpretation of those rules with which elementary memory operates" (D. D. Mordukhay-Boltovskoy).

2. Extension to exceptional cases

Here the matter concerns the application of an actual general rule, but in a special case where some additional facts exclude the possibility of the application of that rule. Most sophisms of this category arise because with an inexperienced calculator the fact escapes notice that the proposed, or obtained, expression is to be subjected to operations which are impossible for the quantities entering its composition.

The learner should be fully aware of the fact that both the forms $0/0$ and $a/0$ are only drawings composed of mathematical symbols and only apparently mathematical expressions, because every mathematical expression loses its sense whenever the divisor becomes equal to zero. Being aware of this fact greatly helps in the analysis of the corresponding mathematical examples. In this connexion they occur not only in books on elementary mathematics but also in textbooks on mathematical analysis.

An example by B. Bolzano (1781–1848):

"If a and b are a pair of different quantities then two identities hold:

$$a - b = a - b,$$
$$b - a = b - a$$

Addition yields:

$$a - a = b - b, \quad \text{or} \quad a \times (1 - 1) = b \times (1 - 1).$$

If we allow the division of both sides of the equality by the factor equal to zero, then we obtain the absurd result: $a = b$ for all a and b." (*Paradoxes of the Infinite* (Paradoksy beskonechnogo), p. 56, Odessa, 1911.) Note that Bolzano does not attempt to divide by zero. Arriving at the expression $a \times 0 = b \times 0$ he points out that division by zero would bring about an absurd conclusion.

The teacher should remember that the "problem consists precisely in teaching the pupils never to make the attempt of dividing by zero" (A. Ya. Khinchin).

3. Ascribing properties of a particular form to the whole species

One often meets errors of this type, consisting in the identification of an arbitrary increasing or decreasing function with a straight line, or with an inversely proportional dependence.

However, the carrying over of the properties of a given form to the whole species may occur in the various problems. This is exemplified beautifully by the famous paradoxes of Ballis, I. Bernoulli, L. Carnot, D'Alembert and many others.

Students are usually interested in geometric embodiments which demonstrate some of these sophisms.

EXAMPLE. In every right-angled triangle the arm is greater than the hypotenuse.

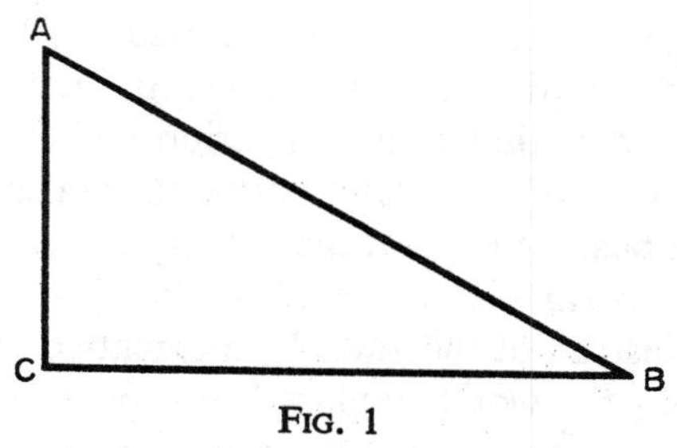

FIG. 1

The difference of the squares of the hypotenuse and one of the arms is $AB^2 - BC^2$ (Fig. 1). This expression may be represented in the form of a product $AB^2 - BC^2 = (AB + BC) \times (AB - BC)$, or $AB^2 - BC^2 = -(AB + BC) \times (BC - AB)$. Dividing right-hand sides by the product $-(AB + BC) \times (AB - BC)$, we obtain the proportion:

$$\frac{AB + BC}{-(AB + BC)} = \frac{BC - AB}{AB - BC}.$$

Since a positive quantity is greater than a negative one, $AB + BC > -(AB + BC)$. But then also $BC - AB > AB - BC$, and therefore $2BC > 2AB$, or $BC > AB$.

Explanation. The assertion that if $a:b = c:d$ and $a > b$, then also $c > d$, holds for positive numbers. It does not extend to a set of numbers which includes negative numbers.

4. Incorrect applications of the principle of direct deductions by the converse

Secondary school students do not immediately understand the need to prove converse theorems. The psychological basis of this phenomenon may be seen in the fact that in a textbook direct theorems are actually followed only by those of the converse theorems which turn out to be correct. Hence in the learner's brain, by association, there enters the false notion of the inevitable correctness of the converse theorem on the strength of the established truth of the direct one. To counteract this impression the teacher should repeatedly give striking examples convincingly demonstrating that such converses are inadmissible.

An even greater difficulty for students is presented by the independent formulation of a conclusion which follows directly from the converse of some affirmative general statement. Frequently on the basis of the fact that "every S is a P," even though that P is not distributed, the students are prone to assert that "all P are S" instead of the partially affirmative "Some P are S." Of course, even the worst pupil will not begin to assert on the basis of the fact that vertically opposite angles are equal, that all equal angles are necessarily vertically opposite. However, even a good pupil still falls into errors of this type when incorrectly formulated converses lead to errors which do not become immediately apparent. On this latter some mathematical sophisms are based.

EXAMPLE 1. Any two numbers are equal.

Say $a \neq b$. Write the identity: $-a = b - (a + b)$ and $-b = a - (a + b)$. Since $(-a) \times b = a \times (-b)$, then

$$\{b - (a + b)\} \times b = \{a - (a + b)\} \times a.$$

Removing the brackets, we have:

$$b^2 - (a + b) \times b = a^2 - (a + b) \times a.$$

Adding $\left(\frac{a+b}{2}\right)^2$ to each member of the equality, we may complete the square of the difference of two numbers:

$$\left(b - \frac{a+b}{2}\right)^2 = \left(a - \frac{a+b}{2}\right)^2.$$

From the equality of the squares of the two numbers we conclude as to the equality of the bases:

$$b - \frac{a+b}{2} = a - \frac{a+b}{2}, \quad \text{where} \quad a = b$$

Explanation. Here the error is introduced in the converse of the statement: "If the bases are equal then their squares are also equal." From this assertion we have directly concluded that "if the squares are equal then also the bases are equal." In fact only a partially affirmative assertion holds: "If the squares are equal then the bases may be equal." This occurs because the statement of the original assertion is not distributed: the squares of not only equal numbers are equal, but also of numbers which are equal only in absolute value.

EXAMPLE 2. In every triangle all the angles are equal.

We shall denote the angles of an arbitrary triangle ABC with unequal sides by α, β, γ, and the sides lying opposite these angles by a, b, c (Fig. 2).

On the extension of the sides BA and CA, lay off the segments AD and AE, equal to b and c respectively. Join the points D and C, E and B.

Since in the $\triangle BEC$, $\angle E = \frac{1}{2}\alpha$, and $\angle CBE = \beta + \frac{1}{2}\alpha$, then according to the law of sines we have:

$$\frac{b + c}{\sin\left(\beta + \frac{\alpha}{2}\right)} = \frac{a}{\sin\frac{\alpha}{2}}. \tag{1}$$

Since in $\triangle BDC$, $\angle D = \frac{1}{2}\alpha$ and $\angle BCD = \gamma + \frac{1}{2}\alpha$, then according to the same law we have:

$$\frac{b+c}{\sin\left(\gamma + \frac{\alpha}{2}\right)} = \frac{a}{\sin\frac{\alpha}{2}}. \tag{2}$$

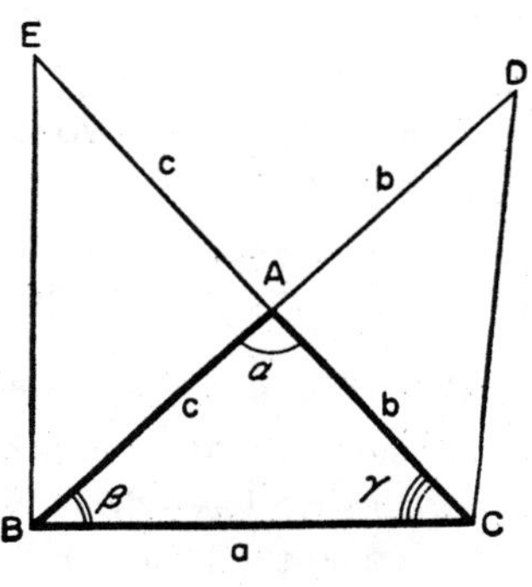

FIG. 2

In the equalities (1) and (2) the right-hand members are equal. Consequently the left-hand members are also equal:

$$\frac{b+c}{\sin\left(\beta + \frac{\alpha}{2}\right)} = \frac{b+c}{\sin\left(\gamma + \frac{\alpha}{2}\right)},$$

where

$$\sin\left(\beta + \frac{\alpha}{2}\right) = \sin\left(\gamma + \frac{\alpha}{2}\right),$$

and therefore

$$\beta = \gamma.$$

Extending the sides AB and CB and repeating analogous arguments, we arrive at the conclusion that $\alpha = \gamma$, which completes the proof of the assertion.

Explanation. On the basis of the fact that

$$\sin\left(\beta + \frac{\alpha}{2}\right) = \sin\left(\gamma + \frac{\alpha}{2}\right),$$

it is possible to make three assumptions:

1. $$\beta + \frac{\alpha}{2} = \gamma + \frac{\alpha}{2} \quad \text{for} \quad \beta = \gamma.$$

2. $$\beta + \frac{\alpha}{2} = 180^\circ \times k - \left(\gamma + \frac{\alpha}{2}\right)$$

for $\alpha + \beta + \gamma = 180^\circ \times k$, where $k = 1, 3, 5, \ldots$

3. $$\beta + \frac{\alpha}{2} = 360^\circ \times k + \left(\gamma + \frac{\alpha}{2}\right)$$

for $\beta - \gamma = 360^\circ \times k$, where $k = 1, 2, 3, \ldots$

The first and the third assumptions should be rejected; one leads to an absurdity, and the other to an impossible requirement, that the difference of two positive angles, each less than 180°, be equal to $360^\circ \times k$, where $k = 1, 2, 3, \ldots$ There remains the second assumption, which for $k = 1$ does not contradict the sense of the problem and relates the angles of the triangle by the known relation.

From this it is apparent, that here also the error is of the type we are now considering. The fact that equal angles have equal sines does not give grounds for concluding the truth of the converse assertion. One may only say: "If the sines of two angles are equal, then also the angles may be equal."

The English mathematician Charles Dodgson (1832–1898) considered that children should become familiar with direct conclusions arrived at by conversion and use examples taken from everyday life long before the time of serious study of mathematics. His children's book, *Alice in Wonderland*, widely known in world literature, was published in England in 1865 under his pseudonym Lewis Carroll, and has had over three hundred editions in that country. In it he gives the following conversation between the heroes of the tale:

"—Do you mean that you think you can find out the answer to it?" said the March Hare.

"Exactly so," said Alice.

"Then you should say what you mean," the March Hare went on.

"I do," Alice hastily replied; "at least I mean what I say—that's the same thing, you know."

"Not the same thing a bit!" said the Hatter. "Why, you might just as well say that 'I see what I eat' is the same thing as 'I eat what I see'!"

"You might just as well say," added the March Hare, "that 'I like what I get' is the same thing as 'I get what I like'!"

"You might just as well say," added the Dormouse, which seemed to be talking in its sleep, "that 'I breathe when I sleep' is the same thing as 'I sleep when I breathe'!"

"It is the same thing with you," said the Hatter, and here the conversation dropped."*

5. Replacement of precise definitions by geometric intuition

The proof of every mathematical argument should be based on the primary concepts, on exact definitions of all remaining concepts, axioms, theorems proved earlier in the given domain of science, and only on them. Definitions eliminate indefiniteness of the concepts (terms) used, which often serves as the cause of various errors. The well-known law of Pascal (1623–1662) requires that, in checking, definitions are to be substituted for terms.

Errors often arise, however, from attempts of students to establish, in the guise of supplementary bases of proof, some data

* Lewis Carroll, *Alice's Adventures in Wonderland*, Macmillan and Co., London, 1953.

of experience deriving from intuitive considerations. This has given cause for the appearance among the analytical minds in the middle of the last century of a tendency to eliminate diagrams from mathematics.

Among beginners the contemplation of a diagram makes a strong impression. It plays the role of an incontestable fact, of which only a suitable explanation is to be sought. Even students under the influence of an intuitive picture are prone to forget the exact definitions of one or another set of concepts, particularly where the visual impression would seem to give completely a direct answer to the problem posed, not requiring an indirect check.

Thus, the problem under consideration is sufficiently complex, and its correct comprehension is of exceptionally great conceptual and educational significance. This is promoted by the analysis of specially chosen examples, constructed on mistaken trust in geometrical intuition, which, it would seem, acts as an equivalent to the corresponding exact definitions. Unfortunately, this applies particularly well to the application of the concept of limit. Sophisms on this subject have found their expression in some collections of mathematical analysis.

EXAMPLE. The sum of the arms is equal to the hypotenuse.

In a right-angled triangle ABC with arms a, b and hypotenuse c, divide the latter into two equal parts (Fig. 3). From the point of division F, drop perpendiculars to the arms. It is easy to see that the length of the broken line $AEFDB$ is equal to the sum of the arms $a + b$. If we now take the points K and L at the mid-points of the segments AF and FB, then by means of the same construction we obtain the broken line $AMKNFPLQB$, whose length is equal to the same sum of the arms $a + b$. Finally, we may consider such a process to be continued indefinitely; however, the length of each of the successively formed broken lines remains unchanged. But, on the other hand, as the number of its links increases, the broken line approaches closer and closer to the hypotenuse of the triangle. In fact, since the length of the segments composing the broken lines decreases without limit, while their ends approach the hypotenuse without limit, it follows that the broken line tends to coincide with the hypotenuse. But then the limit of the length of the broken line will be the length of the hypotenuse. However one and the same quantity cannot have two limits, and therefore it remains to conclude that $c = a + b$.

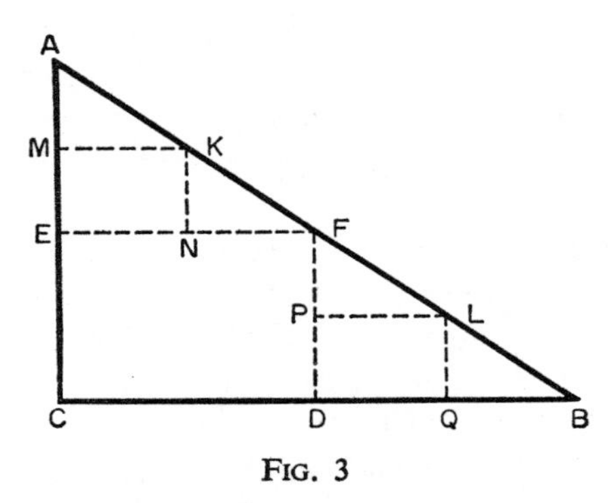

FIG. 3

Explanation. In the above argument an arbitrary conclusion has been introduced: from the tendency of the broken line toward coincidence with the hypotenuse, in the sense cited in the text, there is no basis for concluding that the limit of the length of the broken line is the length of the hypotenuse. Thus, this assumption is unfounded; no foundation may be given for it since it is, in fact, false. Indeed, here we are not within the conditions of applicability of the concept of limit: the difference between a variable quantity, in a special case, a constant (length of the broken line), and its assumed limit (the hypotenuse) is neither an infinitely small quantity nor its special case—zero.

For a better understanding of the problem and to avoid the mechanical carrying-over of this conclusion to the definition of length of a curve, a detailed examination should be carried out.

The eminent champion of popular knowledge, Academician N. A. Morozov (1854–1946) considered that sophisms of the type "the hypotenuse of a right-angled triangle is equal to the sum of its arms" "are of scientific interest, since they draw attention to important peculiarities of mathematical methods, or of mathematical concepts, and of the origin of these concepts in our minds."

6. Errors of construction

We have already considered the problem of errors which may arise in using a correct diagram. Here we shall analyse the basic types of mistakes arising by virtue of a given error in the diagram.

The intention of some mathematicians to annihilate the role of the diagram in geometry somehow could not find a considerable number of followers. And this is fully to be expected, since the whole history of geometry convincingly testifies to the exceptionally great value of the diagram in the discovery of new geometric propositions and the finding of methods for their proof. However, the problem of the mutual relation of logic and intuition in the process of creation and instruction, and the problem of correct use of graphic images, remains up to the present among the pressing problems of methodology of mathematics.

At any rate the students should be well aware that a visual geometrical image still does not guarantee truth, or even logical rigour, and that one should not judge the conditions of a theorem on the basis of the impression given by the diagram. At the same time students should be made aware that a correct diagram, as a rule, helps to make a correct guess and is of great service to apprehension. An incorrect diagram, however, inversely, can be a source of various mistakes. The particular attention to this problem in Soviet mathematico-methodological press of the last few years (the work of Prof. N. F. Chetverukhin, Prof. M. L. Frank, Dr. G. A. Vladimirskiy and others) is explained by its exceptional pedagogic complexity and importance.

In the exposition of geometrical proofs connected with points of intersection of various lines, arguments are usually carried out with reference to a ready-made diagram in which the points of intersection are taken at appropriate places. There is usually no discussion of the question whether the lines under consideration do intersect, and, if they do, where the point of their intersection lies. However, the ability to solve these problems conscientiously and correctly is of essential importance for mastery of methods of geometric proof.

Geometrical sophisms based on errors in construction, among which we distinguish six different forms, serve to point out to the students the value of the requirements just formulated and to develop a critical sense.

Coincident points considered as distinct

EXAMPLE. The exterior angle of a triangle is equal to the interior angle not contiguous to it.

Say, in the quadrilateral $ABCD$, $\angle A$ is supplementary to $\angle C$ (Fig. 4). Since any three points not lying on one straight line completely define a circle, we may therefore assert that through

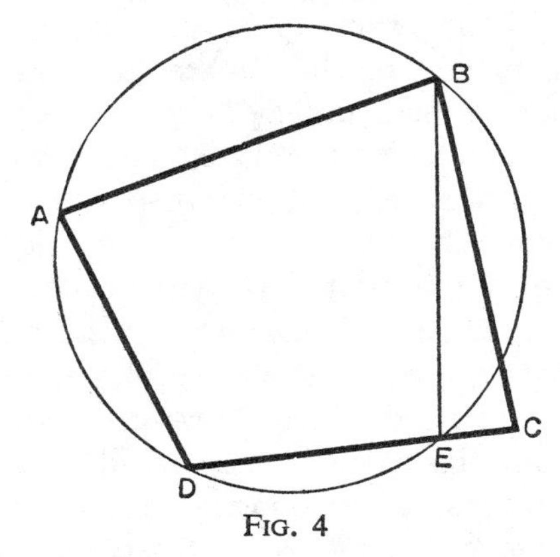

FIG. 4

the points A, B, and D a single circle passes. The point of intersection of that circle with the side DC we denote by E. Joining it by a line segment with the point B we obtain a quadrilateral $ABED$ inscribed in a circle. In it the sum of any two opposite angles constitutes two right angles.

Summing up, we write out two relations:

(1) $\angle A + \angle C = 360°$; (2) $\angle A + \angle BED = 360°$. From them, it is easy to see, it follows that $\angle BED = \angle C$, i.e. the exterior angle of $\triangle BEC$ is equal to the interior angle not contiguous to it.

Explanation. Since in the original quadrilateral the sum of the opposite angles is equal to 360°, all of its vertices, including also C, must lie on the circumference. It follows that the points E and C are not distinct but coincident. By virtue of this the triangle BEC vanishes completely.

Distinct points considered as coincident

EXAMPLE. The area of an equilateral triangle is equal to zero.

In an equilateral triangle ABC construct the altitude AD (Fig. 5). The triangle under consideration is equal to the rectangle

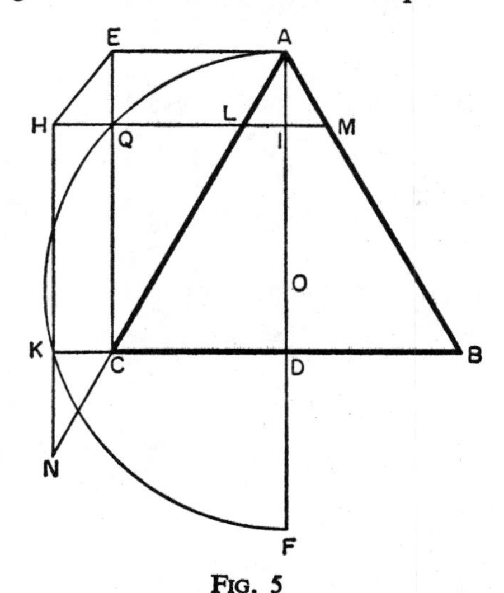

FIG. 5

$ADCE$, the contiguous sides of which are the segments AD and CD. On the extension of AD lay off the segment DF equal to CD. On the segment AF as diameter describe a semicircle to meet the extension of DC at some point K. Then $KD^2 = AD \times DF$. The square $KDIH$, as well as the rectangle $ADCE$, is equal to the triangle ABC.

Representing $\triangle CEA$ as translated along AC in such a way that C coincides with N, E with H and A with L, we note that the square $KHID$ consists of the figure $CLID$ and a figure equal to $MBID$. Hence it follows that the area of the equilateral triangle ALM is equal to zero.

Explanation. The assertion that under the parallel translation the point E coincides with the point H is erroneous. In order to convince ourselves of this, and to be able to estimate the numerical value of the error, we shall find the segments HQ and LI.

$$HQ = KC = KD - CD = \sqrt{(AD \times CD)} - CD$$

$$= \sqrt{\left(\frac{a}{2}\sqrt{(3)} \times \frac{a}{2}\right)} - \frac{a}{2} = \frac{a}{2}\left(\sqrt[4]{(3)} - 1\right)$$

where the letter a denotes the side of $\triangle ABC$.

$$AI = AD - ID = \frac{a}{2}\sqrt{3} - \sqrt{\left(\frac{a}{2}\sqrt{(3)} \times \frac{a}{2}\right)} = \frac{a}{2}\left(\sqrt{(3)} - \sqrt[4]{(3)}\right)$$

$$= \frac{a}{2}\sqrt[4]{(3)}\,(\sqrt[4]{(3)} - 1).$$

$$LI = \frac{AI \times CD}{AD} = \frac{\frac{a}{2}\sqrt[4]{(3)}\,(\sqrt[4]{(3)} - 1) \times \frac{a}{2}}{\frac{a}{2}\sqrt{(3)}}$$

$$= \frac{a}{2}(\sqrt[4]{(3)} - 1) \times \frac{1}{\sqrt[4]{(3)}}.$$

Whence we find:

$$HQ = KC = LI\,.\,\sqrt[4]{(3)} \qquad (\sqrt[4]{(3)} \approx 1{\cdot}3).$$

A point is taken where it cannot lie

EXAMPLE. Every point of the diameter of a circle lies on the circle itself.

Let C be an arbitrary point of the diameter AB. Constructing a fourth harmonic D to the points A, B, C, we divide the segment

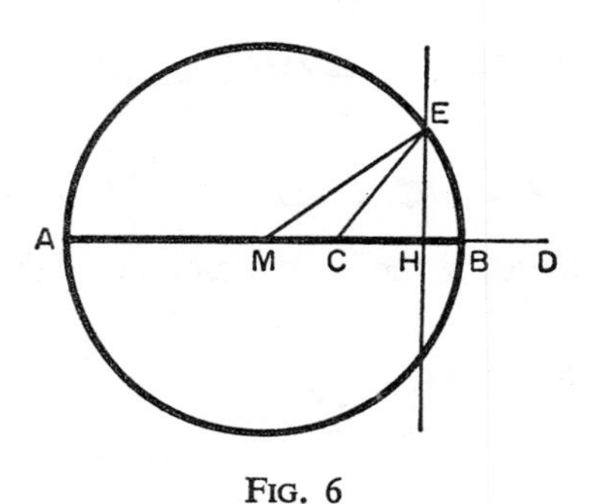

FIG. 6

CD in half, and denote its midpoint as H. If M is the centre of the circle, then, according to a well known theorem, we have $MC \, . \, MD = MA^2$.

We remind the reader of the proof of that theorem. From the diagram we observe directly that the proportion

$$\frac{AC}{CB} = \frac{AD}{DB},$$

which holds for the four harmonic points A, B, C, and D, may be rewritten thus:

$$\frac{MA + MC}{MA - MC} = \frac{MA + MD}{MD - MA}.$$

Having made use of the law of proportion, according to which the sum of the terms of the first ratio is so related to their difference, as the sum of the terms of the second ratio is to their difference, we conclude:

$$\frac{MA}{MC} = \frac{MD}{MA},$$

whence

$$MA^2 = MC \times MD.$$

Noting that $MC = MH - CH$ and $MD = MH + CH$, we establish:

(1) $MH^2 - CH^2 = MC \times MD$ for $MH^2 - CH^2 = MA^2$.

On the other hand, if the perpendicular at the point H to AB intersects the circle at the point E, then

(2) $ME^2 = MH^2 + HE^2$ and
(3) $CE^2 = CH^2 + HE^2$. Subtracting (3) from (2) term by term, we have:
(4) $MH^2 - CH^2 = ME^2 - CE^2 = MA^2 - CE^2$.

But from (1) and (4) it follows that $MA^2 = MA^2 - CE^2$, whence $CE^2 = 0$, i.e. the point C lies on the circle, and since C is an arbitrary point of the diameter, this holds for any point of the diameter AB.

Explanation. In the proportion

$$\frac{AC}{CB} = \frac{AD}{DB}$$

we carry out the transposition of the outside terms:

$$\frac{DB}{CB} = \frac{AD}{AC}.$$

Since $AD > AC$, then, consequently, also DB should be greater than CB. Hence the point H, which should be the midpoint of the segment CD, lies not within the circle but outside it, to the right of the point B.

Thus, the error was the consequence of an incorrect diagram: the point H was taken at a point where it cannot be.

A classical sophism of this type is the "proof" of the proposition that there are no triangles other than isosceles triangles. The Hindu mathematician Sundara Rou gives it as an example of an error, whose very possibility is eliminated by geometrical experiments with a piece of paper. "It would be quite correct," claims Rou, "to ask students to compose these diagrams with a piece of paper. This would give them distinct and exact outlines and would forcefully impress on their minds the truth of the propositions."

The assumed point of intersection is altogether absent

Professor N. F. Chetberukhin in his fundamental work *Chertezhy prostranstvennykh figur v kurse geometrii* (*Diagrams of Three-dimensional Figures in the Geometry Course*), Moscow, 1946, when analysing the pedagogical formulation of the problem of construction of models, emphasizes the following statement: "Particularly great is the importance of representations of spatial figures *in developing space orientation* of the students for a more subtle, more sophisticated spatial thinking, so necessary under the conditions of today's complex technology." A certain role in the solution of this responsible task may be played also by the analysis of stereometrical variants of some sophisms.

EXAMPLE. A right angle is equal to an obtuse angle.

Say, we have a quadrilateral $ABCD$ (Fig. 7), in which the side DA forms a right angle with the side AB and is equal to the side BC, which forms an obtuse angle with the side AB.

From the midpoint of the side AB erect a perpendicular to the plane of the quadrilateral $ABCD$, and from the midpoint of the side DC raise a perpendicular to the straight line intersecting the perpendicular to AB at some point S. Join the point S to the points A, B, C, and D.

From the direct theorem on three perpendiculars it is easy to observe that $AD \perp SA$, i.e. $\angle SAD = 90°$.

Noting that, according to the construction, $AS = SB$ (as inclined lines to a single straight line having equal projections) and $DS = SC$ (as inclined lines to the straight line DC, having equal projections), and from the condition $AD = BC$, we assert that $\triangle SAD = \triangle SBC$.

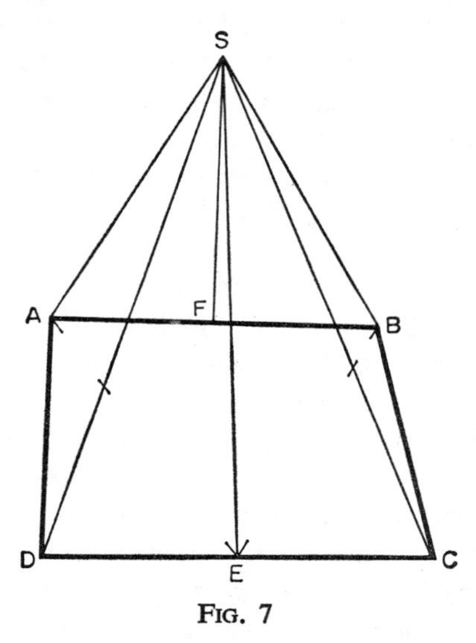

FIG. 7

From the congruence of these triangles we have: $\angle SBC = \angle SAD = 90°$. Applying the inverse theorem on the three perpendiculars, we assert that BC is perpendicular to AB, and therefore $\angle ABC = \angle DAB$, which was to be shown.

Explanation. In the construction used $\triangle DSC$ is isosceles, which leads to a contradiction, since simultaneously with the equality of the inclined lines SD and SC their projections DF and CF turn out to be unequal, which may be seen from considering $\triangle DAF$ and $\triangle CBF$ ($AF = FB, AD = BC$, but $DF < CF$, since $\angle A < \angle B$). Hence the conclusion that the straight lines SF and SE cannot have a common point.

The cause of the very gross error under analysis consists in the fact that in the construction we had postulated the existence of a point of intersection S as the point of intersection of the plane (perpendicular to the given plane raised to it from the midpoint of the segment AB). In short, in the construction we proceeded from the assumption that a plane and a straight line always intersect.

A broken line is taken for a straight line

EXAMPLE. The sophism $64 = 65$ and its generalization.

This sophism is exceptionally valuable in its conceptual significance. The generalized treatment of this sophism, bringing in the use of continued fractions, is set forth by Ignat'yev.* A more elementary exposition of this generalization is given in Chapter IV.

A straight line is taken for a broken line

An error inverse to the preceding.

EXAMPLE. Two triangles are equal if they have two corresponding sides equal and the angles opposite to one of these sides equal.

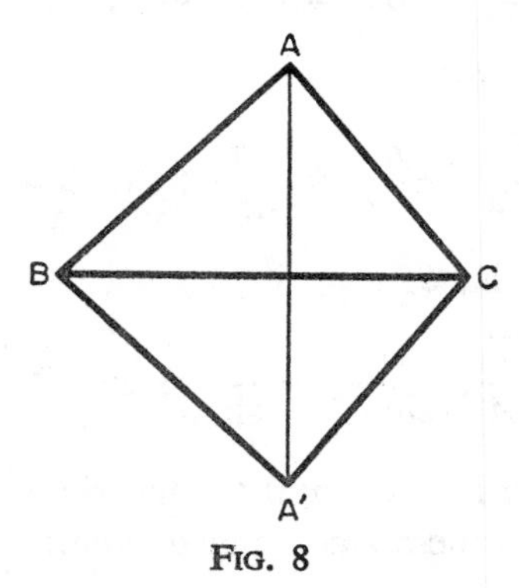

FIG. 8

It is easy to see that this proposition "generalizes" the fourth criterion of equality of triangles. In fact, the usual criterion requires that the angle be taken opposite the greater side. In the formulation given, however, this restriction is taken off.

* E. Ignat'yev, *In the Kingdom of Wit* (V tsarstve smekalki), Book 2, pp. 163–167, St. Petersburg, 1909.

Thus, we are given the triangle ABC and $A'B'C'$, where $BC = B'C'$, $AB = A'B'$, $\angle A = \angle A'$. It is to be shown that $\triangle ABC = \triangle A'B'C'$.

Proof. We lay $\triangle A'B'C'$ and $\triangle ABC$ next to each other in such a way that their equal sides BC and $B'C'$ coincide (Fig. 8), where the point B' coincides with the point B, the point C' with the point C. Join the points A and A'. On the basis of the equality of the segments BA and BA', we conclude that $\triangle BAA'$ is isosceles and, consequently, $\angle BAA' = \angle BA'A$. And since the hypothesis of the theorem gives that $\angle A = \angle A'$, then, dependent on the type of the triangle, by means of addition or subtraction we obtain the relation: $\angle CAA' = \angle CA'A$. This means that $\triangle CAA'$ is isosceles, and therefore $CA = CA'$. The proof is completed by a reference to the third criterion of equality of triangles.

Explanation. By specification we have:

$$AB = A'B' = c; \quad BC = B'C' = a; \quad \angle A = \angle A' = \alpha$$

from $\triangle ABC$ we have:

$$\frac{AB}{\sin C} = \frac{BC}{\sin A}; \quad \frac{c}{\sin C} = \frac{a}{\sin \alpha}; \quad \sin C = \frac{c \times \sin \alpha}{a}. \tag{1}$$

from $\triangle A'B'C'$ we have:

$$\frac{A'B'}{\sin C'} = \frac{B'C'}{\sin A'}; \quad \frac{c}{\sin C'} = \frac{a}{\sin \alpha}; \quad \sin C' = \frac{c \times \sin \alpha}{a}. \tag{2}$$

Since in the relations (1) and (2) the right-hand members are equal, the left-hand members are also equal:

$$\sin C = \sin C'$$

With regard to the angles C and C' three assumptions may be made:

(1) $\angle C = \angle C'$. In this case we have indeed equal triangles. Their congruence is established by means of the first or the second criterion of congruence of triangles.

(2) $\angle C = 180° - \angle C'$. In that case $\angle C + \angle C' = 180°$. Consequently, the diagram given is incorrect: The sides AC and $A'C'$ lie on a single straight line (Fig. 9). A straight line had been taken for a broken line and our whole proof breaks down, as it is based on an incorrect diagram.

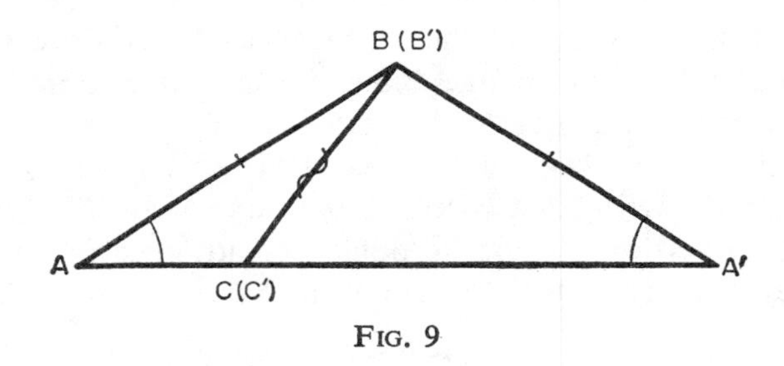

FIG. 9

This is the actual occurrence of the error. However, its cause lies somewhere else—in the incomplete enumeration of all possible cases.

(3) $\angle C = 360° + \angle C'$. This case is impossible, since the difference of two angles of a triangle cannot be equal to 360°.

The refutation of the proposition of the "theorem" is easily attained by constructing triangles satisfying the requirements of the "theorem," but not mutually congruent (see Fig. 45, in subsequent text).

7. Errors which are the consequence of a literal interpretation of the abbreviated conventional formulation of some geometric propositions

Some geometric facts necessary for solving various problems take an abbreviated purely colloquial formulation. Now when it becomes necessary to apply some theorem after a relatively long interval, the essence of the question frequently disappears from memory, and the laconic formulation of the type "two triangles having a side and two angles equal are congruent" may be the cause of various mistakes. A deeper analysis of such mistakes and individual sophisms of the same type have particularly important educational significance.

EXAMPLE. A chord not passing through the centre of a circle is equal to the diameter.

Given is a circle in which a diameter AB is drawn. Taking some point C arbitrarily on the circle, join it to the point A. Designate the midpoint of the chord AC as D and draw through it and point B the chord BE. Now join the points C and E.

Consider the triangles ADB and DCE. They have a side and two angles equal: $AD = DC$, by construction; $\angle B = \angle C$ since they are inscribed angles based on the same arc AE; $\angle ADB = \angle CDE$ as vertically opposite angles. And since in congruent triangles "opposite equal angles lie equal sides," $AB = EC$.

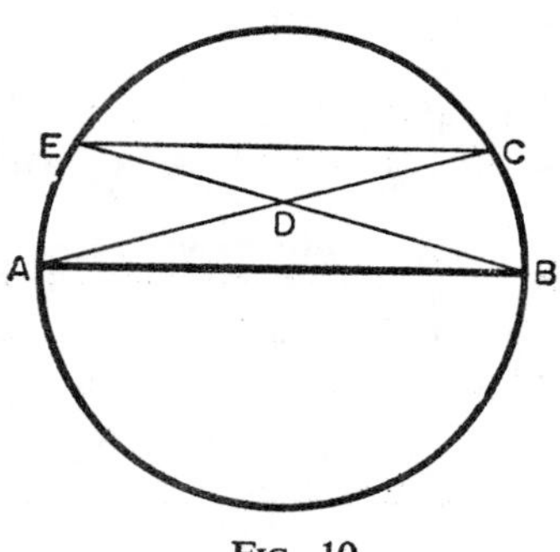

FIG. 10

Explanation. The error was caused by the literal interpretation of a purely colloquial formulation. For comparison we shall recall the correct formulation of the second criterion for congruence of triangles. It asserts: "Two triangles are congruent if a side and its two contiguous angles in one triangle are equal to the side and the two contiguous angles, respectively, of the other triangle." Glancing again at the diagram, we see that we had overlooked the requirement that the equal angles of the triangles be contiguous to their equal sides. In fact, the angles ADB and CDE are constructed correctly, but the angles ABD and DCE—

incorrectly, since the former is not contiguous to the side AD, but lies opposite it.

8. Violation of the sense of conventional notations

The substitution into a symbolic formula of numerical values known not to go beyond the domain of validity is considered as a purely mechanical problem. However, it should not be forgotten that in writing some formulas an element of convention enters. Here an automatic substitution is fraught with absurd conclusions. Thus, if, for example, in the notation $n! = 1 \times 2 \times 3 \times \ldots \times (n-1) \times n$ clarifying the concept of factorial, one sets $n = 3$ and carries out a "thoughtless" substitution, one obtains the following "result": $3! = 1 \times 2 \times 3 \times \ldots \times 2 \times 3$. Students are prone to make a similar error in applying the Newton binomial formula.

EXAMPLE. One is equal to two.

We write the Newton binomial formula:

$$(a+b)^n = a^n + na^{n-1}b + \frac{n(n-1)}{1 \times 2}a^{n-2}b^2 + \ldots + \frac{n(n-1)}{1 \times 2}a^2b^{n-2} + nab^{n-1} + b^n.$$

Since the validity of its use for any natural number n is demontrated in the textbook, there is no hindrance to setting, for example, $n = 1$. The substitution of this value into the Newton binomial formula yields:

$$(a+b)^1 = a + b + 0 + \ldots + 0 + a + b,$$

where $\quad a + b = 2(a+b).$

In the special case when $a + b = 0$, this result does not lead to a contradiction. But in the general case, when $a + b \neq 0$, simplification by that factor leads to the assertion $1 = 2$.

Explanation. In carrying out a "thoughtless" substitution, we forget that the expansion of a binomial has $(n + 1)$ terms, where

n is the exponent of the power of the binomial. The successive writing out of the components of the right-hand member for a given natural number n is terminated by the appearance of the first zero component, representing the $(n + 2)$nd term of the expansion.

9. Deviation from the thesis

Some sophisms are constructed on the basis of the fact that in the course of the proof the absurd thesis of the sophism is replaced by some true assertion. The identification of the true with the false is achieved by the presence of an outward resemblance of their formulations. On the strength of such a substitution the "proof" of a false assertion takes the form of a baseless proof. This will become more vivid if we take, for example, the sophism of Proclus.

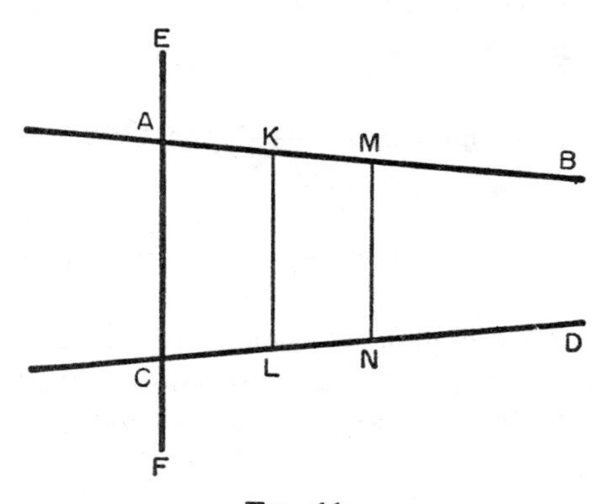

FIG. 11

In his commentaries to Euclid, Proclus tells of the fact that in Greek science there was an attempt to refute the "parallel" postulate. With this aim one tried to show that two lines do not intersect, even when being intersected by a third they form such interior angles that their sum is less than two right angles.

Say two straight lines AB and CD are intersected by a third straight line EF (Fig. 11). On that side of EF for which the sum of the interior angles is less than 180°, lay off on the straight lines AB and CD the segments AK and CL, equal to $\frac{1}{2}AC$. The points K and L cannot coincide, since in that case one would obtain a triangle AKC (or CLA), in which the sum of two sides is equal to the third, which is impossible.

Now join the points K and L by a straight line segment. On the straight lines AB and CD lay off, retaining the direction, the segments KM and LN, equal to $\frac{1}{2}KL$. It is easy to see, as before, that the points M and N cannot coincide.

Since similar arguments may be repeated an arbitrary number of times, we arrive at the conclusion that the straight lines AB and CD do not intersect.

In this proof we meet the substitution of the thesis. In fact, instead of proving the absence of a point of intersection, here one only proves that it cannot in fact be attained by the indicated process of reasoning.

In a simpler case, when the angles CAB and ACD are equal, the argument reduces to a reproduction of the sophism "Achilles and the tortoise," and is refuted in the same fashion.

In characterizing the case when the angles CAB and ACD are not equal, we note that, from the fact that the first segment on AB does not intersect the first segment on CD, the second segment of one straight line does not intersect the second segment of the second straight line, and so on, it would be hasty to conclude that no segment at all of the straight line AB intersects some segment of the straight line CD. In the paradoxical argument one had only admitted just the possibility of the intersection of the nth ($n = 1, 2, 3, \ldots$) segment of the straight line AB with the similarly numbered segment of the straight line CD, while silently excluding every possibility of exhibiting that point as the point of intersection of two non-analogous segments.

We shall grasp the purely logical error under analysis even more vividly if we compare it with the following argument with an analogous structure:

Hypothesis. The father of family X does not know the father of family Y. The mother of the first family does not know the mother of the second family. The only son of one family does not know the only son of the other family.

Conclusion. No member of family X is acquainted with a member of family Y.

Sophisms of this type may be constructed also on the basis of the fact that, in the process of bringing a problem to solution, there is a change not in the method of solution, or at least not only in it, but also in the problem itself.

EXAMPLE. A father at his death, has left to his three sons a bequest in which it is stipulated that the brothers divide among themselves a flock of 17 camels in the following way: to the eldest—one half, to the middle—one third, and to the youngest —one ninth.

The old man's will seemed to be incapable of exact fulfilment, since the brothers did not admit the thought of cutting camels into parts. However, making use of a clever device, they began by borrowing one camel from a neighbour and adding it to their flock. Thereupon the realization of the division did not call forth any more difficulties. As the result of it, the eldest received 9 camels, the middle—6, and the youngest—2. Now there was no more need for the neighbour's camel and it was returned to its owner.

Thus, it follows that the father's will, at least in theory, may be exactly carried out not only in fractions but also in whole numbers.

Explanation. The father had left an improvident will. The sum of the parts $\frac{1}{2} + \frac{1}{3} + \frac{1}{9}$ makes $\frac{17}{18}$ and not unity. The exact fulfilment of the bequest does not agree with the requirements of expediency and is not practicable, as it proposes to give to the eldest $\frac{17}{2}$ heads of cattle, to the middle one $\frac{17}{3}$ and to the youngest $\frac{17}{9}$. Together this constitutes $\frac{289}{18} = 16\frac{1}{18}$, while $\frac{17}{18}$ of one camel remains outside the requirements of the division. The wise advice consisted in such a replacement of the thesis of the will, thanks to which the $\frac{17}{18}$ part was taken not of 17 but of 18 units, which exactly coincides with the number of the flock undergoing division. Such a solution of the problem represents, however, not a precise realization of the will, but only an expedient approximation to the fulfilment of its conditions. In actual fact, the eldest son has received $9 - \frac{17}{2} = \frac{1}{2}$ of a camel more, the middle son received $6 - \frac{17}{3} = \frac{1}{3}$ more, and the youngest, $2 - \frac{17}{9} = \frac{1}{9}$ more. These additions taken together exhaust the remaining $\frac{17}{18}$ of a camel remaining outside the division.

CHAPTER II

Arithmetic

I. Examples of False Arguments

1. The location of thirteen persons, one each, in twelve rooms

A hotel manager succeeded in solving the following, at first sight insoluble, problem: to place in twelve single rooms thirteen persons without putting two persons in one room.

Warning the thirteenth (the man entered under that number in the list of arrivals) that he will be temporarily placed in the first room, the enterprising manager undertook to place the remaining persons one in each room beginning with the first.

At the end of the process two men turned up in the first room; a third man was placed in the second room, a fourth—in the third, a fifth—in the fourth and so on up to the twelfth, who, of course, was placed in the eleventh room.

The twelfth room, which, as we see, remained free, the manager gave to the temporary inhabitant of the first room—the thirteenth client of the hotel.

As result, the problem is solved: $12 = 13$.

2. Eleven fingers on two normal hands

One can perfectly well convince oneself of the presence of ten elements in one set by counting from one to ten by the so-called reverse count. However, if we apply the count to the fingers of two entirely normal hands in the fashion shown on the accompanying illustration, and add up the results, then the number of fingers will turn out to be equal to eleven.

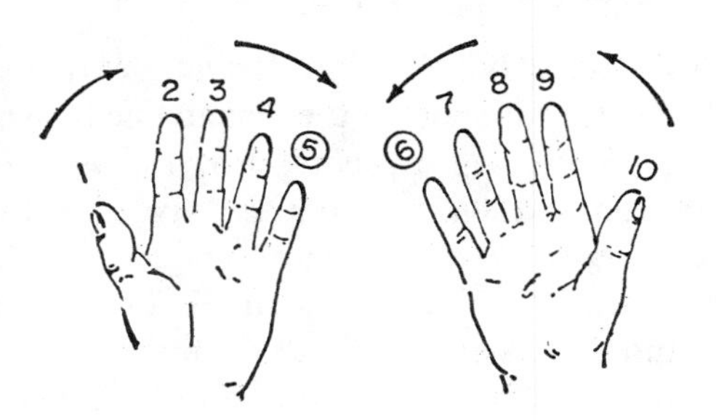

3. Square dollars

As is well known, every set of two equalities may be multiplied term by term. Applying this theorem to the following two equalities:

$$a \text{ dollars} = 100a \text{ cents}$$

$$1 \text{ dollar} = 100 \text{ cents},$$

we obtain a new equality:

$$a \text{ dollars} = (100a \times 100) \text{ cents}$$

or

$$a \text{ dollars} = 10{,}000a \text{ cents},$$

which is clearly untrue.

4. $45 - 45 = 45$

Someone stubbornly asserted that $45 - 45 = 45$.

In support of this, he argued thus:

We write the subtrahend in the form of a sum of the consecutive natural numbers from 1 to 9 and the minuend in the form of the sum of the same numbers, but taken in the opposite order (from 9 to 1).

Locating the subtrahend under the minuend

$$9 + 8 + 7 + 6 + 5 + 4 + 3 + 2 + 1 \qquad (1)$$

$$1 + 2 + 3 + 4 + 5 + 6 + 7 + 8 + 9, \qquad (2)$$

we proceed to the calculation of the difference. With this aim we successively subtract the numbers of the second line from the numbers of the first line, beginning by subtracting the 9. Since 9 cannot be subtracted from 1, then, taking a unit from the two, we have: $11 - 9 = 2$. In a similar way we obtain for the differences of the numbers 11 and 8, 12 and 7, 13 and 6, 14 and 5 respectively, and 3, 5, 7, and 9. Carrying out the subtractions of four from five, three from seven, two from eight, and finally one from nine, we obtain in succession the following results: 1, 4, 6, 8.

Thus:

$$\begin{array}{r} 9 + 8 + 7 + \dot{6} + \dot{5} + \dot{4} + \dot{3} + \dot{2} + 1 \\ - \quad 1 + 2 + 3 + 4 + 5 + 6 + 7 + 8 + 9 \\ \hline 8 + 6 + 4 + 1 + 9 + 7 + 5 + 3 + 2 \end{array}$$

It is not hard to establish, that

$$8 + 6 + 4 + 1 + 9 + 7 + 5 + 3 + 2 = 45.$$

Thus $45 - 45 = 45$.

5. $40 \div 8 = 41$

Little Peter did not like to calculate mentally

In solving the problem of dividing 40 nuts equally among 8 boys, he turned to the scheme of written division

When he carried out the division, it looked thus:

$$\begin{array}{l} 8)\,40\,(41 \\ \underline{32} \\ 8 \\ \underline{8} \\ 0 \end{array}$$

The answer obtained, in fact, worried Peter. He understood very well that each of the boys cannot obtain more nuts than their total number, but nevertheless he could not discover his mistake in the division.

Help little Peter to understand his mistake.

6. Two times two is—five!

As the initial relation we shall take the following obvious equality:

$$4:4 = 5:5. \tag{1}$$

Upon taking out the common factor from each member of the equality (1) we have

$$4 \times (1:1) = 5 \times (1:1)$$

or

$$(2 \times 2) \times (1:1) = 5 \times (1:1). \tag{2}$$

Finally, knowing that $1:1 = 1$, from relation (2) we establish:

$$2 \times 2 = 5. \tag{3}$$

7. Is there a proportionality here?

We consider a few problems together with their answers obtained as a result of applying the simple rule of three.

I. An aeroplane has climbed to an altitude of 8 km in 32 min. To what altitude will it climb in four hours?

Answer. $$\frac{8 \times 60 \times 4}{32} = 60 \text{ (km)}.$$

II. A motor of $1\frac{1}{4}$ h.p. installed on a boat gives the boat a speed of 8 km/hr. What speed is given to the same boat by a 10 h.p. motor?

Answer. $$\frac{8 \times 10}{1{\cdot}25} = 64 \text{ (km/hr)}.$$

III. The winner of a race has run 100 m in 10·2 sec. How far will he run in one hour?

Answer. $$\frac{100 \times 60 \times 60}{10\cdot 2} \approx 35\cdot 3 \text{ (km)}.$$

IV. A boy threw an 800 g discus 12 m high. How far will he throw a discus of 20 g?

Answer. $$\frac{12 \times 800}{20} = 480 \text{ (m)}.$$

V. By means of a stopwatch it had been established that the period of time between the first and the sixth successive ticks of a watch was 6 sec (the stopwatch was switched on at the first tick and stopped at the sixth).

How long will it take the watch to give 12 consecutive ticks?

Answer. $$\frac{6 \times 12}{6} = 12 \text{ (ticks)}.$$

8. 100 per cent economy

On the cover of a technical journal there are a few advertisements. One of them, in recommending some improvement in a steam engine promises a 40 per cent economy of fuel. A second, proposing another improvement, independent of the first, announces an economy of 35 per cent and, finally, a third patented modification, independent of the first two, gives a 25 per cent fuel economy.

Someone, having gained the first improvement, and thereupon the second and, finally, the third one, was very much aggrieved by the fact that the machine, nevertheless, did not work without fuel. He considered himself to be cheated and accused the advertisements of scientific and factual fraud.

"Look," complained the disappointed fellow, "together they promised a 100 per cent economy of fuel, since it is clear to anyone that the sum of 40, 35, and 25 makes 100."

Of course the calculation is incorrect. What fuel economy is actually achieved by the introduction of these three inventions?

9. How should a mean percentage be computed?

We shall consider a somewhat more complicated problem.

A factory has produced the following:

	Total number of articles in tons	Percentage of articles of one type
1st quarter	161 tons	85%
2nd quarter	207 tons	62%
3rd quarter	120 tons	88%
4th quarter	185 tons	86%

Find the mean percentage of articles of one type.

Determining the total number of articles of one type in tons for each quarter we find that during the whole year 673 tons of the article were produced, from which the total number of articles of one type amounts to

$$161 \times 0{\cdot}85 + 207 \times 0{\cdot}62 + 120 \times 0{\cdot}88 + 185 \times 0{\cdot}86 = 529{\cdot}89 \text{ (tons)},$$

which constitutes 78·7 per cent of the total output.

If we simply take the mean number of percentages of the product of quality I by quarters, then we shall obtain

$$(85 + 62 + 88 + 86) : 4 = 80{\cdot}25 \ (\%),$$

which is considerably more than the correct answer.

How do we explain this discrepancy?

10. What will a yearly growth of 40 per cent give in five years?

According to the five-year plan a certain company should increase its production threefold. During the first three years it increased its productivity by 30 per cent yearly (as compared with the preceding year). Is it fulfilling the task of its five-year plan or not?

We recall that productivity of an enterprise is the quantity of production for some definite period of time, for example per day (daily productivity). If one takes the productivity at the beginning of the five-year plan as 100 per cent, then (for a threefold increase during the five-year period) at the end of the five-year period the productivity should become equal to 300 per cent, i.e. increase by 200 per cent. With such an increase of productivity for the five-year period, for one year it should obtain an increase on the average of 200 per cent : 5 = 40 per cent. Consequently, the concern under consideration, in undergoing a yearly growth of 30 per cent is not fulfilling its five-year plan.

However, this conclusion is quite incorrect. In the argument there is a flaw. Find it.

11. A new rule for multiplying fractions

A student disclosed to his mathematics teacher: "I have found a new rule for multiplying mixed fractions, much simpler and easier to apply than the one which you explained to us and the one described in textbooks. The point is that in adding mixed fractions one has to add separately the integers and separately the fractions. For example:

$$6\tfrac{1}{2} + 2\tfrac{1}{4} = (6 + 2) + (\tfrac{1}{2} + \tfrac{1}{4}) = 8 + \tfrac{3}{4} = 8\tfrac{3}{4}.$$

The same is done also in subtraction: from integers we subtract the integers, from the fractions—the fractions, in case of necessity we do a "carry," for example:

$$6\tfrac{1}{2} - 2\tfrac{1}{4} = (6 - 2) + (\tfrac{1}{2} - \tfrac{1}{4}) = 4 + \tfrac{1}{4} = 4\tfrac{1}{4}.$$

$$6\tfrac{1}{2} - 2\tfrac{3}{4} = 5\tfrac{3}{2} - 2\tfrac{3}{4} = (5 - 2) + (\tfrac{3}{2} - \tfrac{3}{4})$$
$$= 3 + \tfrac{3}{4} = 3\tfrac{3}{4}.$$

Obviously one should proceed also in the same way in multiplying mixed fractions: the integer should be multiplied by the integer and the fraction by the fraction, for example:

$$6\tfrac{1}{2} \times 2\tfrac{1}{4} = (6 \times 2) + (\tfrac{1}{2} \times \tfrac{1}{4}) = 12 + \tfrac{1}{8} = 12\tfrac{1}{8}.$$

My rule is simpler to apply and easier to understand than yours."

We wonder whether multiplication of mixed fractions may not be performed in the way proposed by the young inventor?

12. What happened to the dollar?

In a box there were two baskets of pears, 150 in each. The price of the pears was determined in the following somewhat peculiar way: from the first basket the pears are sold at one dollar for ten, and from the second basket at a dollar for fifteen. Thus, for all the pears from the first basket the amount received was $150 : 10 = 15$ (dollars), for all the pears from the second basket $150 : 15 = 10$ (dollars), and in all 25 dollars.

The seller thought that by taking from the first basket ten pears and from the second basket fifteen he should sell 25 pears for 2 dollars. Therefore he mixed the pears from both baskets together and sold the $150 \times 2 = 300$ (pears) at 2 dollars for 25. As a result he received $2 \times (300 : 25) = 24$ (dollars), i.e. one dollar less than the expected sum of the receipts.

What happened to the dollar?

13. Where did the extra dime come from?

In a snack bar there were two baskets of pears of different sorts, 60 pears in each. For this merchandise it was proposed to receive 9 dollars 50 cents, according to the following somewhat unusual calculation: 30 cents for 4 pears from the first basket and 50 cents for 6 pears from the second basket.

However, the waitress, in order to simplify her work mixed together the pears of both kinds and proceeded to sell ten pears of the mixture for 80 cents.

As a result of the sale of pears, there turned out 10 cents too much: not $9.50 but $9.60.

Where did this extra dime come from?

14. A father's will

According to the will of their father three sons were to divide among themselves a herd of seven horses in such a way that the eldest was to receive one half of the herd, the middle one—one fourth, and the youngest—one eighth.

The father's will worried his heirs considerably. In fact, its implementation made it mandatory to cut horses into parts.

However, a way was found from the difficult situation. An old neighbour added his own horse to the herd undergoing division and to the satisfaction of the brothers proceeded with the division.

As a result of it, the eldest son received four horses, the middle one—two, and the youngest—one. The neighbour's horse, not needed any longer, was gratefully returned to its clever owner.

Thus, it turns out that the father's will admits of a solution in integers.

Is it so?

15. $2 \times 3 = 4$

Somebody undertook to demonstrate that 3 times 2 makes not 6 but 4. In carrying out his strange venture, he took in hand an ordinary match and asked those present to follow carefully the process of his thought.

"By breaking the match in halves," stated the strange mathematician, "we shall have once two. Doing the same to one of the halves we shall have a second time two. Finally, carrying out the same operation on the second one of the halves, we shall obtain a third time two. Thus, taking three times two we have obtained four and not six, as one is accustomed to think."

Demonstrate his error to the confused fellow.

II. Analysis of the Examples

1. In the argument a flaw is introduced: the second client of the hotel remained without a room, since we have simply "forgotten" his existence in distributing the rooms of the hotel.

However, here one should not limit oneself to the simple indication of the error. One should explain why this gross mistake is not immediately apparent to everyone.

The explanation of this fact is that the concept of positive whole number is not unambiguous: it may be both a cardinal and an ordinal number. By consciously mixing the concepts of a cardinal and ordinal number one arrives at an illusion of plausibility. In fact we have argued thus: "as result of the distribution in the first room there turned out to be two persons"—a cardinal number; "the third person was located in the second room"—an ordinal number. Such a structure of the argument distracted the attention of the reader from the fact that the second client was left out.

This sophism emphasizes the need to understand with complete clarity what meaning is given to the terms used, particularly when they belong to a set of ambiguous mathematical terms.

2. In the argument there was introduced an intentional identification of the concepts of an ordinal and cardinal number.

The error arises because for the fingers of the right hand the coincidence of the ordinal and the cardinal numbers does not occur.

3. We note that by taking units of length instead of monetary units and proceeding in exactly the same way, we do not obtain anything untrue:

$$a \text{ m} = 100a \text{ cm},$$

$$1 \text{ m} = 100 \text{ cm},$$

$$a \text{ m}^2 = (100a \times 100) \text{ cm}^2,$$

or

$$a \text{ m}^2 = 10{,}000a \text{ cm}^2.$$

As a result of multiplying the numbers expressing metres we obtain a number expressing no longer simple (linear), but square metres, and so we see where the error was introduced in the case of the money: to multiply a dollar by a dollar is not permissible, since "square dollars" and "square cents" do not exist.

Sometimes it is said that the multiplicand may be either an abstract or a concrete number, but the multiplier has to be an abstract number. For example a multiplication such as $2\text{ m} \times 3\text{ m} = 6\text{ m}^2$ shows that this is untrue. One may produce an arbitrary number of other examples of multiplication of a concrete number by a concrete number. Thus, the quantity of work which has to be expended to carry out some task is expressed usually by the number of *workdays*, i.e. the product of the number of workmen by the number of days during which they are occupied. The work of transport is characterized by the number of *passenger-miles* (the product of the number of the passengers by the number of miles). In every instance the multiplication of two concrete numbers leads to some new, *complex* unit. The term by term multiplication of two equalities of concrete numbers is valid whenever there exists a corresponding *complex* unit. Thus, having the equalities

$$1\text{ t} = 1000\text{ kg}, \quad 1\text{ km} = 1000\text{ m},$$

we can multiply them out term by term and obtain the quite correct result:

$$1\text{ t-km} = 1{,}000{,}000\text{ kg-m}.$$

This shows that, in order to raise a weight of one metric ton to a height of one kilometre it is necessary to expend one million times as much work as to raise a weight of one kilogram by one metre.

4. The error consists in the fact that the borrowed unit was given the status of ten.

The question arises: is it by accident that the sum of the differences thus obtained is equal to 45?

No, not by accident. In fact we have "borrowed" a unit five times and each of the borrowed units was considered as a ten.

In this way we formed the sum of five superfluous nines, adding up to 45.

The gross error introduced is not always discovered; it is the result of a false analogy with the process of subtraction of numbers, written out according to the principles of the positional decimal system of counting.

An attempt was made to propose the thesis of this sophism "$45 - 45 = 45$" as a problem to find its "solution": "How is one to subtract 45 (the sum of the numbers one to nine) from 45 in such a way as to obtain as result . . . 45?" (the periodical *Ogonek*, 1953, No. 26, p. 32). Certainly, such a formulation of the problem should call forth a strong protest on the part of the representatives of mathematico-methodological community.

5. Little Peter's big error, as discovered by checking, is not immediately apparent to all children, and many are hard put to explain it. This occurs because of the fact that a correct thought lies at the basis of the little boy's argument. It is that exact division (division without remainder) reduces to the successive subtraction of the divisor from the dividend until a difference is obtained which is equal to zero.

In carrying out the division of 40 by 8, Peter was to establish how many times can 8 be subtracted from 40. He had established that it may be performed four times ($8 \times 4 = 32$) and again once ($8 \times 1 = 8$), i.e. the division is carried out thus:

$$40 : 8 = (32 + 8) : 8 = 32 : 8 + 8 : 8 = 4 + 1.$$

However, since during the subtraction Peter had incorrectly written down the first digit under the column of tens: $4 + 1$ was in his case equated to $4 \times 10 + 1$, i.e. 41. Therein lies his error.

6. By arithmetic it is known that on dividing the terms of a ratio by one and the same number, other than zero, the value of the ratio is not changed. Contrary to this, in the argument under analysis, the value of the ratio (in the given case equal to unity) was multiplied by a number which is the greatest common divisor of the terms of the ratio.

In the argument an illusion of plausibility was given on the basis of a false analogy with the distributive property of multiplication with respect to addition.

It is essential to note that such errors are impossible, if one uses the fraction line as the sign for division. Indeed

$$4 : 4 = \tfrac{4}{4} = 4 \times \tfrac{1}{4} = 4 \times (1 : 4)$$

and

$$5 : 5 = \tfrac{5}{5} = 5 \times \tfrac{1}{5} = 5 \times (1 : 5).$$

7. The incorrectness of the answers obtained is immediately apparent (with the exception of the fifth answer, whose falsity is not so obvious). They are all obtained on the assumption that among the two quantities concerned in the problem there exists a direct or an inverse proportionality. In fact, however, the dependences between the quantities with which we are dealing in each of these problems is more complicated.

In solving the first problem we arrive at the conclusion that in four hours the aeroplane would climb to 60 km. This would be true if the altitude increased in exact proportion to time. In fact, however, such a proportionality does not exist: as the altitude of the climb rises, the time necessary for climbing each additional metre also rises, and every aeroplane has its "ceiling," i.e. a climbing altitude that it cannot exceed.

In the second problem the solution is given on the assumption that the speed of the motor boat is proportional to the horsepower of the motor. It appeared that a relatively small motor, 10 h.p. in all, gave the boat a velocity of 64 km/hr, i.e. the velocity of a mail train (in Russia). This calculation is also incorrect, since in fact the increase of velocity of a motor boat, just as of any other vessel with mechanical traction, occurs much more slowly than the increase of the power of the motor. Experiments show that the power increases approximately as the cube of the velocity: in order to increase the velocity a times, it is necessary to increase the motive power not a times, but $a \times a \times a = a^3$ times. In order to give the boat a velocity of 64 km/hr, i.e.

increase its velocity 8 times as compared to that already indicated in the conditions of the problem (8 km/hr), the power of the motor should be increased $8 \times 8 \times 8 = 512$ times, i.e. the motor of 1·25 h.p. has to be replaced by a motor of $1{\cdot}25 \times 512 = 640$ h.p. The replacement of a 1·25 h.p. motor by a 10 h.p. motor, i.e. by a motor 8 times more powerful, will give only a doubling of the velocity, since $2 \times 2 \times 2 = 8$, and the motor boat instead of 8 km/hr will go at $8 \times 2 = 16$ (km/hr).

The solution of the third problem states that a man traversing 100 m in 10·2 sec will run over 35 km in an hour. As experience shows, a man's strength in the course of such a fast run is exhausted very quickly, and in each succeeding 100 m the man will run much slower, and then will stop altogether. There can be no question of proportionality between the duration of the run begun at such a high speed and the distance traversed.

The fourth problem is solved on the assumption that a man throws a discus a distance inversely proportional to the weight of the discus. The fantastic answer obtained (that the discus is thrown 480 m (almost half a kilometre) shows the incorrectness of this assumption. This is explained by the fact that it is only with very small variations of the weight of the discus being thrown, that the distance traversed by it varies approximately with inverse proportionality to the weight (to be more exact: if one throws a projectile in a vacuum at the same angle to the horizontal, and transmits to it the same quantity of energy, then in order to increase the distance traversed by it 2, 3, 4, in general a times, it is necessary to decrease the weight of the projectile $2 \times 2 = 4$ times, $3 \times 3 = 9$ times, $4 \times 4 = 16$ times, in general $a \times a = a^2$ times). When the discus is thrown in the air with a decrease of the weight of the discus and increase of its velocity ever greater and greater, then the air resistance presents an ever greater influence on the distance of its flight. Besides, when a discus is of very small weight and is thrown by hand it is impossible to impart quite the same quantity of energy as that which can be imparted to a heavier discus. For these reasons, very small discuses fly not further, but less far than heavier ones, and the calculation of the distance of flight based on the application of an inversely proportional dependence yields results which are quite incorrect.

In solving the fifth problem, we proceeded from the assumption that the number of ticks and the time are in a directly proportional ratio. However, these two quantities are mutually related by another, much more complicated dependence.

This is clear from the following considerations.

Since any two successive ticks are separated by a single time period, it is clear that six ticks following one after another are mutually separated by five periods. Thus, in six seconds 6 ticks occur, mutually separated by five periods, the duration of each constituting 1·2 seconds.

Our assertion of the absence of proportionality follows from the fact that, for twelve ticks there are not ten, but eleven periods between them.

Now the answer to the problem: "How long will it take for the watch to give 12 ticks?" no longer presents any difficulty.

Obviously, the watch will give 12 ticks in $(12 + t)$ seconds, where by the letter t we have denoted the length in seconds of the period of time between two successive ticks ($t = 1{\cdot}2$ sec).

From all the above, the resolution should be made to observe a high degree of caution in solving problems by the rule of three: in every instance, before applying that rule one should convince oneself that the quantities under consideration in the problem are indeed in a proportional relationship. An indefinite number of examples may be given, based on the application of the rule of three in cases when it cannot be applied. We shall meet more of them.

8. Here the error is introduced by calculating the percentage of economy of the fuel. In the presence of all three improvements the calculations of the economy of fuel should be carried out in the following way: 40 per cent of the fuel economy obtained from the introduction of the first invention should be taken from the entire quantity of the fuel used by the machine; 35 per cent of the economy from the introduction of the second invention is subtracted from the remainder after the subtraction of the economy by virtue of the first improvement; 25 per cent of the fuel economy, obtained from the introduction of the third improvement, is subtracted from the second remainder. The calculation will be correct also if we carry out the calculation in another order, i.e. begin from the second or the third improvement. Let us take a specific example. Assume, that the machine requires 100 kg of fuel. Then the computation should be carried out in the following fashion:

40% of 100 equals 40	100 – 40 = 60 (kg)
35% „ 60 „ 21	60 – 21 = 39 „
25% „ 39 „ 9·75	39 – 9·75 = 29·25 (kg).

Consequently, with the three improvements, the boiler will take instead of 100 kg, only 29·25 kg of fuel.

The result may be expressed thus:

$$100 \times (1 - 0{\cdot}40) \times (1 - 0{\cdot}35) \times (1 - 0{\cdot}25) = 29{\cdot}25 \text{ (kg)}.$$

This expression shows that the total percentage of economy is independent of the order in which we carry out the separate improvements.

In the erroneous solution of this problem we meet the use of a false analogy.

Since in the solution of numerous problems we are accustomed to taking the initial quantity as 100 per cent (for example in the problem under consideration: "if one takes the quantity of fuel used by the machine as 100 per cent"), the students quite often are liable to assume that percentages are always to be subtracted from the initial number indicated in the problem ("the fundamental 100 per cent quantity"), and often overlook the necessity of taking into account the variation of that quantity (the new, other "fundamental, 100 per cent quantity") as result of the defined operations (in the given case as result of a successive introduction of improvements giving an economy of fuel).

9. It is necessary to remember strictly that in calculating the mean percentage x of several groups (or, in the parlance of statisticians, of several subsets) the simple arithmetic mean of the numbers $p_1, p_2, p_3, \ldots, p_n$, expressing the respective percentages of every group separately, yields a correct result only in the case when all these groups are of equal quantity (have the same weight). If among these groups not all of them are of equal quantity, the mean percentage x has to be computed according to the formula of the weighted mean, namely:

$$x = (p_1a_1 + p_2a_2 + p_3a_3 \ldots + p_na_n) : (a_1 + a_2 + \ldots + a_n),$$

where $a_1, a_2, a_3, \ldots, a_n$ are numbers expressing the quantity (weight) of each group. In the case $a_1 = a_2 = a_3 = \ldots = a_n$ this formula reduces to the formula of the simple, i.e. unweighted arithmetic mean $x = (p_1 + p_2 + p_3 + \ldots + p_n) : n$.

10. In order to recognize the error we shall begin by contrasting the correct argument with the false one.

Say we have a yearly growth of 40 per cent in comparison with the preceding year. At the beginning of the first year the productivity was 100 per cent, at its end it was already 100% + 40% = 140%. The new 40 per cent increment for the second year should be computed not for the initial productivity of 100 per cent, but for that which it was at the beginning of the second year, i.e. for 140 per cent. Consequently, the growth for the second year is equal to 40 per cent of 140 per cent, i.e. 56 per cent, and at the end of the second year the productivity will constitute 140% + 56% = 196%. Adding here 40 per cent of 196 per cent, we find that at the end of the third year we have already 196% + 78·4% = 274·4%. At the end of the fourth year we have 274·4 per cent plus 40 per cent of 174·4 per cent, or 274·4% + 109·76% = 384·16%, and at the end of the fifth year 386·16 per cent plus 40 per cent of 384·16 per cent, or 384·16% + 153·66% = 537·82%.

Thus, the yearly growth of productivity of 40 per cent yields not a three-fold but a more than five-fold growth of productivity for the five-year period.

In order to have a three-fold growth in the five-year period we should take yearly not a 40 per cent growth, but less. Taking 20 per cent, we find that at the end of the first, second, etc., years of the five-year period we have 120 per cent, 144 per cent, 172·8 per cent, 207·36 per cent, 248·83 per cent, and a three-fold increase in the five-year period will not be reached. With a 25 per cent yearly growth, a three-fold growth for the five-year period is already ensured—at the end of the five-year period we obtain 305·18 per cent of the productivity. A yearly growth of 30 per cent will give 371·29 per cent at the end of the five-year period.

The error in the first argument, which led to the conclusion that a yearly growth of 30 per cent would not yield a three-fold growth for the five-year period, was that the percentage of the percentage (cumulative percentages) had not been taken into account: 30 per cent of the productivity for the year should not be considered from the productivity at the beginning of the five-year period taken by us as 100 per cent, but from the productivity at the beginning of each year. In other words, we are dealing not with simple, but with compound percentages. A more precise calculation, based on the application of the formula for compound interest:

$$A = a\left(1 + \frac{p}{100}\right)^n$$

(for solving this problem take $A = 3a$, $n = 5$, and carry out the calculation of p by means of logarithms), leads to the conclusion that the yearly growth of $p = 24{\cdot}6$ per cent already ensures a three-fold increase for the five-year period (300·32 per cent), and a yearly increase of $p = 24{\cdot}5$ per cent does not ensure it (299·12 per cent).

11. We shall consider the problem: if a man traverses $6\frac{1}{2}$ km in an hour, then how much will he traverse in $2\frac{1}{4}$ hr? We solve this problem without making use of any rule of multiplications of fractions, neither the old nor the new.

In an hour the man will walk $6\frac{1}{2}$ km or 6500 m, in 2 hr $6500 \times 2 = 13{,}000$ (m), in a quarter of an hour $6500 : 4 = 1625$ (m), and in all in the $2\frac{1}{4}$ hr $13{,}000 + 1625 = 14{,}625$ (m), or 14 km 625 m, or $14\frac{5}{8}$ km.

It would seem that this problem can be solved in one operation—the multiplication of $6\frac{1}{2}$ by $2\frac{1}{4}$. Indeed, for any whole number of hours the path traversed is equal to the path traversed in one hour, repeated as many times as the number of hours the motion continued, i.e. is equal to the product of $6\frac{1}{2}$ and the number of hours. It is natural to expect that also with a fractional number of hours the result should be obtained by means of the same operation of multiplication. But if one adopts the new rule for multiplication, then the product of $6\frac{1}{2}$ and $2\frac{1}{4}$, as we had seen above, turns out to be equal to $12\frac{1}{8}$ and not $14\frac{5}{8}$, as it should be. The usual rule for multiplication in the present case gives:

$$6\tfrac{1}{2} \times 2\tfrac{1}{4} = \frac{13 \times 9}{2 \times 4} = \frac{117}{8} = 14\tfrac{5}{8},$$

i.e. just that result which we have obtained above, without using fractions (by means of dividing kilometres into metres).

Thus, the "new rule" of multiplication of mixed numbers has to be rejected. But it is interesting to clarify why the addition (and subtraction) of mixed numbers may be carried out by adding (and subtracting) separately the whole and separately the fractional numbers, while multiplication cannot be carried out in this fashion (i.e. by multiplying whole number by whole number and fraction by fraction).

In order to add a third number to the sum of two numbers, it is necessary to add that number to one of the components, leaving the other unchanged, e.g. in order to add 2 to $8 + 5$, we have to take either $(8 + 2) + 5 = 10 + 5$, or $8 + (5 + 2) = 8 + 7$. In both cases we obtain the correct result (15). Here we have the formula:

$$(a + b) + c = a + (b + c),$$

expressing the so-called "associative" property of addition. Applying the "commutative" property of addition ($a + b = b + a$), we obtain $(a + b) + c = (b + a) + c = b + (a + c) = (a + c) + b$ and arrive at another expression of the "associative property" of addition:

$$(a + b) + c = (a + c) + b.$$

In order to multiply the product of two numbers by a third number, it is necessary to multiply by this third number one of the factors, leaving the second one unchanged. For example, in order to multiply 3×4 by 5 one should take either $(3 \times 5) \times 4 = 15 \times 4$, or $3 \times (4 \times 5) = 3 \times 20$. In both cases we obtain the correct result (60). Here we have the formula:

$$(a \times b) \times c = a \times (b \times c),$$

expressing the "associative" property of multiplication.

We have considered the case when the same operation is carried out twice: two times an addition, or two times a multiplication. Now we shall take a case in which first an addition and then a multiplication is carried out. In order to multiply the sum of two numbers by a third number, it appears that we must not limit ourselves to the multiplication of only one of the given numbers: we have to multiply each of the components. For example, in order to multiply by 5 the sum $3 + 4$, we must not take $3 \times 5 + 4 = 15 + 4 = 19$ or $3 + 4 \times 5 = 3 + 20 = 23$, but it is necessary to take $3 \times 5 + 4 \times 5 = 15 + 20 = 35$. Thus, we must not limit ourselves to the simple association of the third number with one of the first two. Here the second operation (multiplication) is as if distributed among the two numbers, over which the first operation is carried out (addition). One says that the product of a sum possesses the distributive property, which is expressed by the formula:

$$(a + b) \times c = ac + bc.$$

Thus, when adding to the sum of two numbers a third number, we have to apply the associative property, and when multiplying the sum of two numbers by a third number we have to apply the distributive property.

We now return to operations upon fractions. Every mixed number may be considered as a sum of two numbers, an integer and a proper fraction; for example, $6\frac{1}{2} = 6 + \frac{1}{2}$. Adding two mixed numbers, for example $6\frac{1}{2}$ and $2\frac{1}{4}$, we can add to $6\frac{1}{2}$ first 2, and then $\frac{1}{4}$. In order to add to $6\frac{1}{2} = 6 + \frac{1}{2}$ the number 2, we increase by 2 the first component (6), and the second ($\frac{1}{2}$) we leave without change. We obtain $(6 + 2) + \frac{1}{2}$. To this sum of two components $(6 + 2)$ and $\frac{1}{2}$ it remains to add still $\frac{1}{4}$. Applying once again the associative property, we leave the first component $(6 + 2)$ unchanged, and the second ($\frac{1}{2}$) we increase by $\frac{1}{4}$. Finally, we have:

$$6\tfrac{1}{2} + 2\tfrac{1}{4} = (6 + 2) + (\tfrac{1}{2} + \tfrac{1}{4}) = 8 + \tfrac{3}{4} = 8\tfrac{3}{4}.$$

Now consider the multiplication of the mixed numbers $6\frac{1}{2}$ and $2\frac{1}{4}$.

In order to multiply $6\frac{1}{2} = 6 + \frac{1}{2}$ by $2\frac{1}{4}$, applying the distributive property, we have to multiply by $2\frac{1}{4}$ both the first component (6), and the second component ($\frac{1}{2}$) and then add the products:

$$6\tfrac{1}{2} \times 2\tfrac{1}{4} = (6 + \tfrac{1}{2}) \times 2\tfrac{1}{4} = 6 \times 2\tfrac{1}{4} + \tfrac{1}{2} \times 2\tfrac{1}{4}.$$

But to obtain the product $6 \times 2\frac{1}{4}$ or $2\frac{1}{4} \times 6$ or $(2 + \frac{1}{4}) \times 6$, we again make use of the distributive property:

$$6 \times 2\tfrac{1}{4} = (2 + \tfrac{1}{4}) \times 6 = 2 \times 6 + \tfrac{1}{4} \times 6 = 6 \times 2 + 6 \times \tfrac{1}{4}.$$

In precisely the same way we proceed for obtaining the product $\frac{1}{2} \times 2\frac{1}{4}$, namely:

$$\tfrac{1}{2} \times 2\tfrac{1}{4} = (2 + \tfrac{1}{4}) \times \tfrac{1}{2} = 2 \times \tfrac{1}{2} + \tfrac{1}{4} \times \tfrac{1}{2} = \tfrac{1}{2} \times 2 + \tfrac{1}{2} \times \tfrac{1}{4}.$$

Finally, it turns out that our product of two mixed numbers is equal to the sum of the four partial products, namely:

$$6\tfrac{1}{2} \times 2\tfrac{1}{4} = 6 \times 2 + 6 \times \tfrac{1}{4} + \tfrac{1}{2} \times 2 + \tfrac{1}{2} \times \tfrac{1}{4}.$$

Now it is clear that, by multiplying the integer by the integer and the fraction by the fraction, we obtain not the whole product, but only a portion of it: the third and the fourth partial products are lost ($6 \times \frac{1}{4}$ and $\frac{1}{2} \times 2$), i.e. the product of the integer part of the first factor with the fractional part of the second factor, and of the integer part of the second factor with the fractional part of the first.

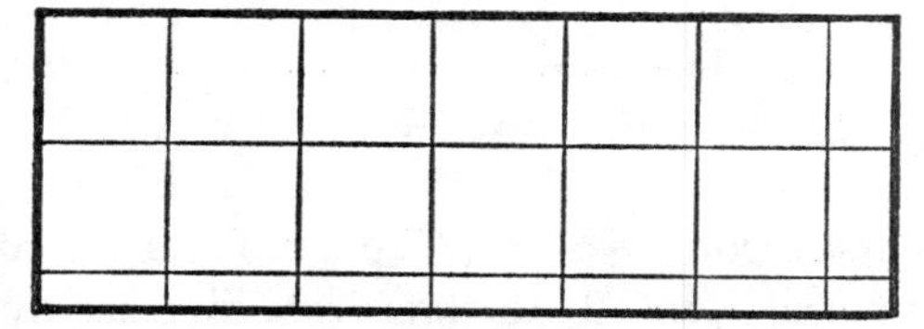

FIG. 12

Using multiplication to compute the area of a rectangle from its length and width, we can represent quite clearly all the four partial products entering the product of two mixed numbers. Taking, for example, a rectangle with sides $6\frac{1}{2}$ and $2\frac{1}{4}$, we see (Fig. 12), that its area consists of four parts: a large rectangle with sides 6 and 2, a narrow long strip—the rectangle with sides 6 and $\frac{1}{4}$, a somewhat wider strip on the right—a rectangle with sides $\frac{1}{2}$ and 2, and a minute rectangle with sides $\frac{1}{2}$ and $\frac{1}{4}$. The area of the large rectangle contains 12 squares (square units), the area of the long narrow strip—six quarters of a square, or $1\frac{2}{4} = 1\frac{1}{2}$ square, the area of the strip on the right—two halves, or one whole square, and the area of the minute rectangle is one-half of a quarter, or one-eighth of a square. In all we have $12 + 1\frac{1}{2} + 1 + \frac{1}{8} = 14\frac{5}{8}$, as it should be.

If we denote the integral and the fractional parts of both factors by letters, then we obtain just the rule for multiplication of binomials, well known from the algebra course:

$$(a + b)(c + d) = ac + bc + ad + bd,$$

graphically represented in Fig. 13.

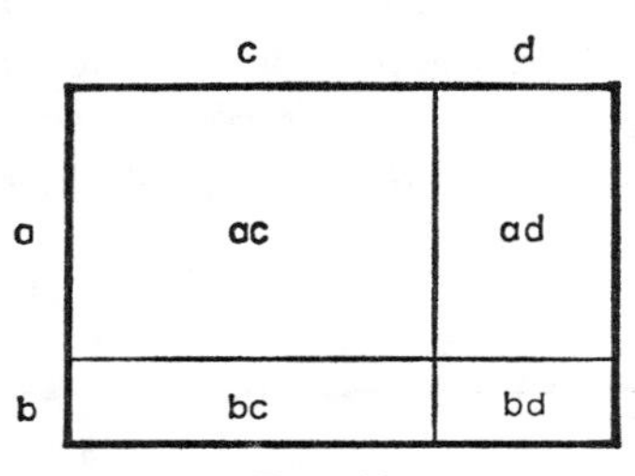

FIG. 13

12. From the second basket (15 pears for a dollar) the seller could take 15 pieces, in all, 10 times. Adding to every 15 pears ten pieces each from the first basket (10 pears for a dollar), he took out from it only 100 pears. The ten sets of pears thus composed he sells for 20 dollars. After this there will remain 50 pears only from the first basket. Every ten of these pears he was supposed to sell at a dollar for ten and obtain 5 dollars for the 50 pears. In fact, however, he sold these 50 pears for 4 dollars (at two dollars for 25) and thus suffered a loss of one dollar.

In explaining this sophism one should remind students of the correct calculation of the cost of 25 pears in the case of selling a mixture.

Since the cost of the 150 pears in the first basket constitutes 15 dollars, and the cost of 150 pears in the second basket is 10 dollars, the cost of each pear of the mixture will be $\frac{15 + 10}{150 + 150} = \frac{25}{300} = \frac{1}{21}$ of a dollar. Consequently, the cost of 25 pears constitutes $\frac{1}{12} \times 25 = \frac{25}{12}$ dollars. In all, it is possible to obtain the following number of sets of 25 pears from the set of pears in the two baskets: $300:25 = 12$. In selling them, one will obtain $\frac{25}{12} \times 12 = 25$ (dollars), i.e. the same amount as in the separate sale at different prices.

In the argument under analysis we have met with a deviation from the thesis: under the guise of changing the manner of solution, the problem is replaced by another not equivalent to the first.

The problem equivalent to the initial one is formulated in the following way: "Every set of 25 pears, composed of 10 pears from the first basket and of 15 pears from the second is sold at 2 dollars. The remainder of the pears in the first basket is sold at the price established earlier."

In the erroneous argument the original problem is replaced by the following one, not equivalent to it: "it had been decided to mix 150 pears of one sort, costing 15 dollars, and 150 pears of another sort, costing 10 dollars, and to sell them at two dollars for 25 pieces. What is the sum to be received by selling the pears at the indicated price (compare with the previously established cost of this merchandise)?"

13. In this argument the same error is introduced as in the argument of Problem 12.

It is meant for an independent verification by students, to which it becomes comprehensible on the basis of the preliminary analysis of the similar argument under the leadership of the teacher.

14. In the will of the deceased parent an impossibility is introduced. In fact, the sum of the shares $\frac{1}{2} + \frac{1}{4} + \frac{1}{8}$ constitutes $\frac{7}{8}$ and not unity. To carry out the will exactly would require the eldest son to receive $\frac{7}{2}$ heads of the herd, the middle one $\frac{7}{4}$ and the youngest $\frac{7}{8}$. Altogether this constitutes $\frac{49}{8} = 6\frac{1}{8}$, and $\frac{7}{8}$ of a horse remains outside the requirements of the division.

The old neighbour by his assistance has prompted a replacement of the thesis of the will, such that one took $\frac{7}{8}$ not of 7, but of 8 units. However, such a solution of the problem realized the will only inexactly, since the eldest son has received $4 - \frac{7}{2} = \frac{1}{2}$

horses more, the middle one $2 - \frac{7}{4} = \frac{1}{4}$ horses more, and the youngest $1 - \frac{7}{8} = \frac{1}{8}$ of a horse more. These additions, when grouped together, exhaust the $\frac{7}{8}$ of a horse which had been left out of the repartition.

15. In the argument a deviation from the thesis had been introduced: the problem of the number constituting the product of two units by three is substituted for by the problem of the number of pieces of the match, obtained as result of the process defined, which was used as a false illustration of the argument under analysis.

We call the illustration false, because in it the role of multiplicand is at first played by the two halves of a whole match, and thereupon by two quarters of the same match.

CHAPTER III

Algebra

I. Examples of False Reasoning

16. Half a dollar is equal to five cents

Everyone will of course agree that

$$\tfrac{1}{4} \text{ dollar} = 25 \text{ cents}. \qquad (1)$$

However, taking the square root of both members of the equality (1), we obtain:

$$\tfrac{1}{2} \text{ dollar} = 5 \text{ cents}, \qquad (2)$$

i.e. half a dollar is equal to five cents.

17. 6 = 2

The following equation was proposed for solution:

$$\sqrt{(x)} + x = 2. \qquad (1)$$

Quickly and confidently a student wrote the following computation in his copybook:

$$\sqrt{(x)} = 2 - x; \quad x = 4 - 4x + x^2;$$
$$x^2 - 5x + 4 = 0 \qquad (2)$$
$$x_1 = 4; \; x_2 = 1.$$

Being firmly convinced of the correctness of his transformations and calculations, the student was very much discouraged by the results of the substitution of the value of x_1 in eqn (1). It seemed to him that he had demonstrated something patently absurd: $6 = 2$.

Help the student to understand his error.

18. 12 = 6 = 0

In the copybook of a young man studying mathematics, a curious entry was found, which we reproduce in its entirety:

"Solve the equation:

$$3\sqrt{(x)} + x + 2 = 0.$$

Solution.

$$3\sqrt{(x)} = -x - 2 \quad (1); \quad 9x = x^2 + 4x + 4;$$

$$x^2 - 5x + 4 = 1.$$

$$x_1 = 4; \qquad x_2 = 1.$$

Check. For $x = 4$ we have:

$$3\sqrt{(4)} + 4 + 2 = 12, \text{ i.e. } 12 = 0. \quad (!)$$

For $x = 1$ we have:

$$3 + 1 + 2 = 6, \text{ i.e. } 6 = 0. \quad (!)$$

Answer."

Here the entry of the solution of the equation was interrupted. Help the young man to overcome his difficulties.

19. Divisibility of polynomials and divisibility of numbers

Having established that some assertion about a property of numbers is true in a series of special cases, e.g. for a series of definite numbers, we nevertheless cannot be assured that this assertion will turn out to be true always: to conclude *from the particular to the general* one may only make an assumption, a guess, which has to be either proved or refuted in what follows. But if such a proof is carried out, then we may boldly apply our general proposition to every special case: the conclusion from the general to the particular is completely valid. Thus, by subtracting unity from the squares of prime numbers greater than three, namely the numbers 5, 7, 11, 13, 17, etc., we obtain the numbers 24, 48, 120, 168, 288, etc.—multiples of 24. It is natural to make the guess: are not multiples of 24 all numbers of the form $p^2 - 1$, where p ($p > 3$) is a prime number? Representing $p^2 - 1$ in the form $(p + 1) \times (p - 1)$, we note that the numbers $p + 1$ and $p - 1$ are both even, and one of them is even doubly even, i.e. is divisible not only by two but by four, and therefore $p^2 - 1$ is a multiple of 8. On the other hand, of three consecutive integers $p - 1$, p, $p + 1$, one is invariably a multiple of three, and as p, being a prime number greater than 3, cannot be a multiple of three, then one of the numbers $p - 1$ or $p + 1$ has to be the multiple of three, and consequently so also is their product. Thus, whatever the prime number $p > 3$, $p^2 - 1$ is always a multiple of $8 \times 3 = 24$. Having proved this assertion in its general form, we may be assured that it will turn out to be correct also in every particular case, i.e. for every number satisfying the conditions imposed.

We shall assume, however, that some assertion proven in its general form turns out to be untrue in some particular case. It is clear that here there invariably occurs some confusion or simply a mistake: either the general assertion which we had considered as proven is actually untrue, i.e. an error had been introduced into the proof, or else the statement that in the given case our general assertion is not fulfilled, is itself false.

We have a curious example of this kind of confusion in the following argument.

Take the binomial $x^n - a^n$, where n is an arbitrary natural (i.e. positive whole) number; x and a are arbitrary, mutually unequal real numbers, where $a \neq 0$. Setting $x/a = y$, we have:

$$x = ay,$$

$$x^n - a^n = a^n(y^n - 1),$$

$$x - a = a(y - 1),$$

$$a^{n-1}(x - a) = a^n(y - 1).$$

Noting that a^n is divisible by a, and $y^n - 1$ is divisible by $y - 1$ (the difference of the powers is always divisible by the difference of the bases), we conclude that

$$x^n - a^n \text{ is divisible by } a^{n-1}(x - a).$$

This is the general, and proved, assertion. We apply it now to the particular case $a = 2$, $x = 3$, $n = 3$. We have

$$x^n - a^n = 3^3 - 2^3 = 19, \quad a^{n-1}(x - a) = 2^2(3 - 2) = 4,$$

and, consequently, 19 is divisible by 4!!!

Account for this.

20. An arbitrary number *a* is equal to zero

Say, a is an arbitrary real number distinct from zero.

Set up the quadratic equation:

$$x^2 - ax = -\tfrac{1}{3}a^2. \tag{1}$$

Solving this equation within the set of real numbers, a student argued thus:

Multiply both members by $-3a$, and then add the difference $x^3 - a^3$ to both members. We obtain:

$$-3ax^2 + 3a^2x = a^3$$

$$x^3 - 3ax^2 + 3a^2x - a^3 = x^3.$$

Making use of the formula for the cube of a difference of two numbers, we rewrite the equation in a shorter form:

$$(x - a)^3 = x^3,$$

and upon taking the cube root of both members we have:

$$x - a = x, \tag{2}$$

whence $a = 0$. Thus, it has turned out, that an arbitrary real number a, distinct from zero, is equal to zero.

21. 7 = 13

The equation

$$\frac{x+5}{x-7} - 5 = \frac{4x-40}{13-x} \qquad (1)$$

may be transformed thus:

$$\frac{x+5-5(x-7)}{x-7} = \frac{4x-40}{13-x};$$

$$-\frac{4x-40}{x-7} = \frac{4x-40}{13-x}; \qquad \frac{4x-40}{7-x} = \frac{4x-40}{13-x}. \qquad (2)$$

From relation (2) we draw the conclusion, that 7 = 13.

22. Positive unity is equal to negative unity

Say b is a positive number other than one.
We define the number a in such a way that

$$b^a = -1. \qquad (1)$$

Proceeding from this relation, we assert, that $b^{2a} = 1$.
It is easy to see that $a = 0$, since, according to the condition, $b \neq 1$.
From the same conclusion it follows that $b^a = 1$.
Comparing the relations (1) and (2), we establish that

$$1 = -1.$$

23. Another "proof" of the equality of the positive and negative unities

From the algebra course of grade X it is known that an imaginary unity is designated by the letter i, and that the square of that imaginary unity is equal to -1, i.e. $i^2 = -1$.

However, on the other hand,

$$i = \sqrt{(-1)} \text{ and } \sqrt{(-1)} \times \sqrt{(-1)}$$
$$= \sqrt{((-1) \times (-1))} = \sqrt{(1)} = 1,$$

and therefore, obviously, $i^2 = 1$.

Thus, it follows that negative unity is equal to positive unity.

24. The imaginary unit and the negative real unit are equal

Say $x = -1$. Where $x^4 = (-1)^4 = +1$, and as $+1 = (-1) \times (-1)$, then

$$x^4 = (-1) \times (-1). \qquad (1)$$

Extracting from both members the fourth root, we obtain

$$x = \sqrt[4]{((-1) \times (-1))} = \sqrt{[\sqrt{((-1) \times (-1))}]}$$
$$= \sqrt{[\sqrt{(-1)} \times \sqrt{(-1)}]} = \sqrt{(i \times i)} = \sqrt{(i^2)} = i, \qquad (2)$$

where the letter i denotes, as is usual, the imaginary unit, i.e. the square root of the number -1 ($i = \sqrt{(-1)}$). Carrying out the transformations indicated in line (2), we have made use of the following algebraic theorems: (1) the radical of a radical may be replaced by a single radical whose exponent is equal to the product of the exponents of the given radicals. On the basis of the converse theorem we have replaced a fourth root by the square root of a square root of the expression under the radical sign; (2) the root of the product is equal to the product of the roots of the factors; (3) the operations of extracting a square root and that of squaring mutually cancel each other out.

In line (2) we have established that $x = i$. But we proceeded from the fact that $x = -1$. Consequently, $i = -1$, i.e. the imaginary unit and the negative real unit are equal.

What error in the argument has led us to such an absurdity?

25. $i^2 = 1$

The equality

$$\frac{1}{-1} = \frac{-1}{1}$$

calls forth no doubts.

Applying the operation of extracting the square root, we have:

$$\frac{\sqrt{(1)}}{\sqrt{(-1)}} = \frac{\sqrt{(-1)}}{\sqrt{(1)}},$$

i.e.

$$\frac{1}{i} = \frac{i}{1},$$

whence

$$i^2 = 1.$$

26. Every negative number is greater than the positive number having the same absolute value

The following argument is based on an obvious fact: if the first term of the first ratio of some proportion is greater than its second term, then also the first term of the second ratio of that proportion is greater than its second term. In short, if

$$a > b \text{ and } a : b = c : d, \text{ then } c > d.$$

We take an arbitrary positive number a and, noting that

$$(+a) : (-a) = -1$$

and

$$(-a) : (+a) = -1,$$

we set up the proportion:

$$(+a) : (-a) = (-a) : (+a).$$

Here the first term of the first ratio is greater than its second term, since $+a > -a$. According to the above, in such a case also the first term of the second ratio is greater than its second term. Consequently, we obtain that $-a > +a$.

27. If $a > b$, then $a > 2b$

Take two arbitrary positive numbers a and b, and assume that $a > b$. Multiplying both members of this inequality by b, we obtain a new inequality $ab > b^2$. Subtracting a^2 from both members, we arrive at the inequality $ab - a^2 > b^2 - a^2$, or, what is equivalent, the inequality:

$$a(b - a) > (b + a)(b - a). \qquad (1)$$

Upon division of both members by $b - a$ we have the relation:

$$a > b + a. \qquad (2)$$

If to this last inequality we add, term by term, the original inequality $a > b$, then we obtain the inequality $2a > 2b + a$, or, upon subtracting a from both members, the inequality:

$$a > 2b. \qquad (3)$$

Thus, if $a > b$, then $a > 2b$. For example, from the fact that $10 > 9$, we conclude according to what we have just demonstrated, that $10 > 18$.

28. If a and b are positive numbers, then $a > b$ and $b > a$

As is well known, if we have two inequalities in the same sense, i.e. both with the sign $>$ (greater than) or both with the symbol $<$ (less than), we may add or multiply them, term by term, and the new inequality will have the same direction as the two given ones: from the inequality $a > b$ and $c > d$ follows that

$$a + c > b + d, \quad ac > bd.$$

Take two positive numbers a and b and write the following two quite indisputable inequalities:

$$a > -b, \quad b > -b.$$

Multiplying them term by term, we arrive at the conclusion that $ab > b^2$, or, after dividing both members by b $(b > 0)$,

$$a > b.$$

Now if we write two other, just as indisputable, inequalities:

$$b > -a, \ a > -a,$$

then it will turn out that $ba > a^2$ and $b > a$.

Thus, of any two positive numbers each is greater than the other.

29. A positive number is less than zero

Say a and b are arbitrary positive numbers satisfying the inequality:

$$a > b. \tag{1}$$

Multiplying both members of the relation (1) by $b - a$, we have:

$$a(b - a) > b(b - a).$$

The further transformations require no explanation:

$$ab - a^2 > b^2 - ab;$$

$$0 > a^2 - 2ab + b^2;$$

$$0 > (a - b)^2. \tag{2}$$

However $(a - b)^2$, where $a \neq b$, is a positive number, since the square of every real number, other than zero, is positive.

Thus, the relation (2) allows us to assert that a positive number is less than zero.

30. The sum of natural numbers, each of which exceeds unity, is greater than their product

We begin from the unquestionable equality

$$\frac{a + b + c + \ldots + l}{abc \ldots l} = \frac{a + b + c + \ldots + l}{abc \ldots l}. \tag{1}$$

Taking the logarithms of both members of the equality (1), we write:

$$\log \frac{a + b + c + \ldots + l}{abc \ldots l} = \log \frac{a + b + c + \ldots + l}{abc \ldots l}. \tag{2}$$

From relation (2) we go on to the inequality

$$2 \log \frac{a+b+c+\ldots+l}{abc\ldots l} > \log \frac{a+b+c+\ldots+l}{abc\ldots l}, \quad (3)$$

which we rewrite in the following form:

$$\log \left(\frac{a+b+c+\ldots+l}{abc\ldots l}\right)^2 > \log \frac{a+b+c+\ldots+l}{abc\ldots l}.$$

Exponentiating both members, we have:

$$\left(\frac{a+b+c+\ldots+l}{abc\ldots l}\right)^2 > \frac{a+b+c+\ldots+l}{abc\ldots l}. \quad (4)$$

Finally, dividing both members of inequality (4) by

$$\frac{a+b+c+\ldots+l}{abc\ldots l},$$

we obtain:

$$\frac{a+b+c+\ldots+l}{abc\ldots l} > 1,$$

whence

$$a+b+c+\ldots+l > abc\ldots l.$$

31. What is the square root of the number a^2?

Some students were given a problem: to find the numerical value of the expression $2x + \sqrt{(1 - 2x + x^2)}$ for $x = 3$. This easy problem was quickly solved by the children, but a strange thing happened: some directly substituted the given number 3 for x and, carrying the operations on numbers, arrived at the answer 8, while others, noting that the expression under the radical sign is a square of the difference of the numbers 1 and x, first transformed the given expression to the form

$$2x + \sqrt{(1 - x^2)} = 2x + (1 - x) = 1 + x$$

and upon substituting obtained another answer: $1 + 3 = 4$.

32. Another "proof" that an arbitrary number is equal to zero

Say b is an arbitrary number other than zero.

If
$$a = -b, \tag{1}$$
then, obviously,
$$a^2 = b^2. \tag{2}$$

Taking the logarithms of the relation (2), we shall have
$$2 \log a = 2 \log b. \tag{3}$$

Introducing the new notation:
$$\log a = \alpha, \quad \log b = \beta, \tag{4}$$
write:
$$10^\alpha = a; \quad 10^\beta = b.$$

On the basis of the relations (3) and (4) we assert that $\alpha = \beta$ and therefore:
$$10^\alpha = 10^\beta, \text{ i.e. } a = b. \tag{5}$$

Adding the equalities (1) and (5) term by term, we have:
$$2a = 0, \text{ i.e. } a = 0.$$

Hence we conclude that an arbitrary number, other than zero, is equal to zero.

33. A number does not change if in it arbitrary digits are transposed

In order to prove the "theorem" enunciated in the title of the present section we argue in the following way:

Every number the sum of whose digits is equal to zero is necessarily itself equal to zero, since all of its digits are zero.

Take two arbitrary numbers of at least three digits:
$$N = 100a + 10b + c,$$
$$N_1 = 100a_1 + 10b_1 + c_1,$$
where the letters a, b, c, a_1, b_1, c_1 denote arbitrary digits. The first number has the sum of digits $s = a + b + c$, and the second $s_1 = a_1 + b_1 + c_1$.

Consider the difference

$$N - N_1 = 100(a - a_1) + 10(b - b_1) + (c - c_1).$$

The sum of the digits of this last number $N - N_1$ is equal to

$$\begin{aligned}(a - a_1) + (b - b_1) + (c - c_1) &= a - a_1 + b - b_1 + c - c_1 \\ &= (a + b + c) - (a_1 + b_1 + c_1) \\ &= s - s_1.\end{aligned}$$

If the numbers N and N_1 have the same sums of digits, i.e. if $s = s_1$, then the sum of the digits of the number $N - N_1$ is equal to zero.

Consequently, for $s = s_1$ we always have $N = N_1$.

The numbers N and N_1, differing only by the order of the digits, have one and the same sum of digits. Consequently, such numbers are always equal to each other (for example, $257 = 725$).

The "proof" carried out for three-digit numbers is applicable without essential modification to any multidigit number.

34. What is the meaning of the theorem on the existence of a root in the algebra of complex numbers?

A man heard that there is in higher algebra a theorem that every equation has at least one root, real or imaginary. The man was very much confused upon meeting the irrational equation:

$$+\sqrt{(x)} = -1, \qquad (1)$$

which, upon being freed of the radical, leads to the equation $x = 1$. Having "solved" this latter equation he found its unique root 1. Remembering that a term by term raising of the equation to a power may give extraneous roots, he carried the substitution of the root found in the given equation and found that the root did not satisfy the given equation. Consequently, the given equation had no roots at all. But what happened to the theorem on the existence of a root?

We note that it is possible to indicate an arbitrary number of equations which have no roots at all, similar to the eqn (1). Such, for example, is the equation:

$$\sqrt{(5 + x)} + \sqrt{(5 - x)} = \sqrt{(x)},$$

since both values $x_1 = 0$ and $x_2 = 4$, obtained on solving this equation by freeing it of radicals, do not satisfy it.

35. On a method of obtaining correct results, the application of which requires considerable caution

As is well known, every irreducible fraction m/n in which the denominator n contains at least one prime factor, other than 2 or 5, when transformed into a decimal fraction, gives an infinite periodic fraction. We set ourselves the opposite problem: having some infinite periodic fraction, to find that ordinary fraction from which it had been obtained by transformation into a decimal fraction. Here we shall limit ourselves to "pure" periodic fractions, i.e. to those whose period begins immediately after the decimal sign.

Say, for example, that we are given the fraction 0·242424 . . ., i.e. a pure periodic fraction with the integral part zero and periodic part 24. Denoting the value of this fraction by x, take the equality

$$x = 0{\cdot}242424\ldots$$

and multiply both of its members by 100. We obtain

$$100x = 24{\cdot}242424\ldots = 24 + 0{\cdot}242424\ldots$$

Now this latter equality may be rewritten in a new finite form (the infinite repetition of the period is excluded):

$$100x = 24 + x.$$

We have obtained a first degree equation, whose solution yields $x = \frac{24}{99} = \frac{8}{33}$. To check, we turn $\frac{8}{33}$ into a decimal fraction and upon dividing 8 by 33 we indeed obtain the given periodic fraction 0·242424 . . .

In this way it is possible to transform into an ordinary fraction any pure periodic fraction, but one should not always multiply by 100: if the period contains k digits, then the fraction should be multiplied by 10^k. In order to transform into an ordinary fraction a "mixed" periodic fraction, i.e. one in which other digits intervene between the decimal sign and the period, we must first transform the mixed fraction in such a way as to reduce the problem to the transformation of a pure periodic fraction. For example, if the fraction is $x = 0{\cdot}8333\ldots$, then we first multiply both members of this equality by 10 and then, having

$$10x = 8{\cdot}33333\ldots = 8 + 0{\cdot}333\ldots,$$

we set $y = 0{\cdot}333\ldots$ and find $y = \frac{1}{3}$ and $x = (8 + \frac{1}{3}) \div 10 = \frac{5}{6}$. A check (dividing 5 by 6) shows that the problem is solved correctly.

A characteristic peculiarity of the method was the exclusion of an infinite repetition of the period.

Here is another problem where an analogous method is applied. We are to find $\sqrt{((2) + \sqrt{((2) + \sqrt{((2) + \ldots}}}$, assuming that the operation of extracting the square root is repeated an infinite number of times. Setting $x = \sqrt{((2) + \sqrt{((2) + \sqrt{((2) + \ldots}}}$, we square both sides of this equality and find that, on transferring the number 2 from the left to the right-hand member, we again obtain the expression denoted by x. Carrying out the substitution, we eliminate the infinite set of roots and arrive at a quadratic equation, from which we then determine x:

$$x^2 = 2 + \sqrt{((2) + \sqrt{((2) + \sqrt{((2) + \ldots}}},$$

$$x^2 - 2 = \sqrt{((2) + \sqrt{((2) + \sqrt{((2) + \ldots}}},$$

$$x^2 - 2 = x, \quad x^2 - x - 2 = 0,$$

$$x_1 = 0{\cdot}5 + \sqrt{(2{\cdot}25)} = 0{\cdot}5 + 1{\cdot}5 = 2, \quad x_2 = 0{\cdot}5 - 1{\cdot}5 = -1.$$

Of course, only the positive root is applicable. Thus, the required expression is equal to 2.

The correctness of our assertion is easily checked by carrying out the following calculations:

$$x_1 = \sqrt{(2)} = 1{\cdot}414;$$
$$x_2 = \sqrt{((2) + \sqrt{(2)})} = \sqrt{(2 + x_1)} = \sqrt{(3{\cdot}414)} = 1{\cdot}848;$$
$$x_3 = \sqrt{(2 + x_2)} = 1{\cdot}961;$$
$$x_4 = \sqrt{(2 + x_3)} = 1{\cdot}990;$$
$$x_5 = \sqrt{(2 + x_4)} = 1{\cdot}997;$$
$$x_6 = \sqrt{(2 + x_5)} = 1{\cdot}999.$$

As we see, the sequence of numbers

$$x_1 = \sqrt{(2)}, \quad x_2 = \sqrt{((2) + \sqrt{(2)})},$$
$$x_3 = \sqrt{((2) + \sqrt{((2) + \sqrt{((2)}, \ldots,}$$

which may be arbitrarily extended, consists of numbers which indeed approach the value 2 as its limit (insofar as one may judge from our short calculation carried out with an accuracy of three decimal digits).

Our method for eliminating an infinite series of repeating operations in the two cases considered has led to correct results. But let us consider one more application of this method.

Take some arbitrary positive number a and denote by the letter x the sum of the infinite series whose terms are all equal to a. Then we carry out the elimination of the infinite number of terms and arrive at the unexpected conclusion that $a = 0$:

$$x = a + a + a + \ldots,$$
$$x = a + (a + a + \ldots),$$
$$x = a + x,$$
$$0 = a.$$

The application of the method for eliminating the infinity of terms has led here to the contradiction between the conclusion ($a = 0$) and the original condition ($a > 0$).

Here is another example: denoting by x the value of the algebraic sum of the infinite set of terms

$$1 - 2 + 4 - 8 + 16 - 32 + \ldots,$$

we have:

$$x = 1 - 2 + 4 - 8 + 16 - 32 + \ldots,$$

$$x = 1 - 2 \times (1 - 2 + 4 - 8 + 16 - \ldots),$$

$$x = 1 - 2x, \quad x + 2x = 1, \quad 3x = 1, \quad x = \tfrac{1}{3}.$$

The answer is clearly incorrect, since, finding the sums of the successively increasing number of terms, we obtain a sequence of integers:

$$x_1 = 1, \quad x_2 = 1 - 2 = -1,$$

$$x_3 = 1 - 2 + 4 = 3,$$

$$x_4 = 1 - 2 + 4 - 8 = -5,$$

$$x_5 = 1 - 2 + 4 - 8 + 16 = 11 \text{ and so on.}$$

which show no approach to the number $\frac{1}{3}$. Thus, the method for eliminating an infinite set of terms sometimes produces correct results and sometimes yields results which are clearly incorrect. Hence its application calls for caution: it is necessary to clarify under what conditions this method yields correct results.

36. About the sum $1 - 1 + 1 - 1 + \ldots$

Consider the algebraic sum of an infinite set of terms equal alternately to plus one and minus one. We shall try to find the value of that sum. Denoting it by x, we have:

$$x = 1 - 1 + 1 - 1 + 1 - \ldots \qquad (1)$$

Rewriting the equality (1) in a somewhat different form, namely:

$$x = 1 - (1 - 1 + 1 - 1 + 1 - \ldots),$$

we note that in the brackets we have again obtained the original sum. Replacing it by x, we have the expression $x = 1 - x$, whose root is equal to 0·5.

If we seek the value x of the sum (1) under investigation, having previously included in brackets every pair of components consisting of one positive and one negative term, we obtain

$$x = (1 - 1) + (1 - 1) + (1 - 1) + \ldots,$$
$$x = 0 + 0 + 0 + \ldots,$$
$$x = 0.$$

But we may proceed in another way. We collect the terms in pairs, beginning not with the first, but with the second term, and put a minus sign in front of every pair of terms. Then we obtain

$$x = 1 - (1 - 1) - (1 - 1) - \ldots,$$
$$x = 1 - 0 - 0 - \ldots,$$
$$x = 1.$$

Finally, transposing every positive term in the place of the negative and vice versa, we arrive at the sum:

$$x = -1 + 1 - 1 + 1 - 1 + 1 - 1 + \ldots,$$
$$x = -1 + (1 - 1) + (1 - 1) + (1 - 1) + \ldots,$$
$$x = -1 + 0 + 0 + 0 + \ldots,$$
$$x = -1.$$

Thus, proceeding by four different and, it would seem, equally correct methods, we have arrived at four different conclusions as to the value of x, namely: x turned out to be equal to 0·5, 0, +1, −1. This may be considered as the "proof" of the obvious absurdity, that

$$0{\cdot}5 = 0 = +1 = -1.$$

37. Is the whole always greater than its part?

Consider the set N of all natural numbers, i.e. the totality of all the numbers 1, 2, 3, 4, . . ., etc., and the set Q of the squares of all integers, i.e. the totality of all the numbers $1^2 = 1$, $2^2 = 4$, $3^2 = 9$, $4^2 = 16$, $5^2 = 25$ and so on. (We shall refer to the numbers 1, 4, 9, 16, 25 and so on as *square* numbers.) Every natural number is an "element" of the set N, every square number is an "element" of the set Q. Each of these sets is infinite, but it is clear that the second set is only a part of the first one: for among the natural numbers 1, 2, 3, 4, . . . we meet all the square numbers without exception: 1, 4, 9, 16, . . ., but, besides, also an arbitrary series of numbers that are not squares, such as 2, 3, 5, 6, 7, and so on.

Now write down the natural numbers in increasing order in one line and under each natural number write its square. We obtain two infinitely long lines, whose beginnings are:

$$1, 2, 3, 4, 5, 6, 7, 8, 9, 10, 11, 12, \ldots,$$
$$1, 4, 9, 16, 25, 36, 49, 64, 81, 100, 121, 144, \ldots$$

Comparison of these two lines leads to a quite unexpected conclusion: both of the sets taken, namely the set of all natural numbers (the first line) and the set of all square numbers (the second line) have an equal number of elements. In other words: there are as many square numbers as there are natural numbers, and inversely. Consequently a part (the set Q) is equal to its whole (the set N).

An inequality is some relation between quantities. It should naturally possess certain general properties. One of these properties is expressed by the eighth axiom of book one of Euclid's *Elements*: "The whole is greater than the part."*

The question arises: Is the whole always greater than its part?

38. Another "proof" of the equality of two arbitrary numbers

Take two arbitrary numbers a and $b > a$, and write the identity:

$$a^2 - 2ab + b^2 = b^2 - 2ab + a^2, \qquad (1)$$

where the algebraic sums in the right- and the left-hand members differ from one another only by the order of the terms.

* Euclid, *Elements*, Vol. I, p. 232, Dover, N.Y., 1956.

The equality (1) we rewrite in a shorter form, making use of the formula for the square of a difference:

$$(a - b)^2 = (b - a)^2. \tag{2}$$

Extracting the square root from both members, we obtain:

$$a - b = b - a, \tag{3}$$

whence, upon transferring some terms, simplifying and dividing both members by 2, we have:

$$a + a = b + b, \quad 2a = 2b, \quad a = b. \tag{4}$$

39. The sum of two arbitrary equal numbers is equal to zero

Take an arbitrary number, other than zero, and write down the equality

$$x = a.$$

Now carry out the following steps:

$$-4ax = -4a^2,$$

$$-4ax + 4a^2 = 0,$$

$$x^2 - 4ax + 4a^2 = x^2,$$

$$(x - 2a)^2 = x^2, \tag{1}$$

$$x - 2a = x. \tag{2}$$

Replacing in the last equality x by the number a equal to it, we obtain:

$$a - 2a = a,$$

$$-a = a,$$

$$0 = a + a.$$

40. A number does not change if 1 is added to it

Take an arbitrary number n and proceed from the identity:

$$n^2 - n(2n + 1) = (n + 1)^2 - (n + 1)(2n + 1),$$

whose truth is easy to check by removing the brackets. Adding $\left(\frac{2n+1}{2}\right)^2$ to both members of this identity, we rewrite it in the form:

$$n^2 - 2n \times \frac{2n+1}{2} + \left(\frac{2n+1}{2}\right)^2$$
$$= (n+1)^2 - 2(n+1) \times \frac{2n+1}{2} + \left(\frac{2n+1}{2}\right)^2,$$

or

$$\left(n - \frac{2n+1}{2}\right)^2 = \left(n + 1 - \frac{2n+1}{2}\right)^2, \qquad (1)$$

whence it follows, that

$$n - \frac{2n+1}{2} = n + 1 - \frac{2n+1}{2}, \qquad (2)$$

or

$$n = n + 1.$$

41. Achilles and the tortoise

On a horizontal straight line mark off two points A and B at a distance 100 m apart. (A is on the left and B is on the right.) We assume that along that straight line two points M and N are moving, both from the left to the right, but the point M with a velocity of 10 m/sec and the point N with a velocity of only 1 m/sec (both velocities are considered as constant).

We shall demonstrate that M, chasing N, will never reach it.

When M reaches the point B, it will not find N there: the latter will already be ahead at some point B_1. When M reaches the point B_1, N will again no longer be there, as it will have gone on to some new point B_2, situated to the right of B_1. When M reaches B_2, N will be at a point B_3 situated even more to the right. This argument may be repeated an arbitrary number of times, and therefore M will never reach N, even though it moves ten times faster.

What lapse in this apparently correct reasoning has led to such an absurdity?

The sophism just formulated was pointed out by the Greek philosopher Zeno as early as the fifth century B.C. Zeno was demonstrating that no matter how fast Achilles (a legendary Greek hero) ran, he could never overtake a slowly moving tortoise.

42. On some student's errors

To conclude the present chapter, concerned with the consideration of errors in algebraic arguments, we shall analyse two very simple, but regrettably, very frequent errors, and then we propose a few problems for independent consideration by the reader.

The first refers to the simplification of algebraic fractions: students cross out the same letters in the numerator and the denominator without bothering to note whether these letters denote factors of the entire numerator and the entire denominator or not. Thus, they carry out the simplification of the fraction $\frac{a^2x}{b^2 + cx}$ by x and obtain the fraction $\frac{a^2}{b^2 + c}$, forgetting that to simplify a fraction means to divide its numerator and denominator by one and the same number, and that, in order to divide a monomial numerator a^2x by x, it is sufficient to cross out the x in it, whereas to divide the binomial denominator $b^2 + cx$ by x it is necessary to divide every term, both b^2 and cx by x. Upon division by x the given fraction will take the following form:

$$\frac{a^2}{\frac{b^2}{x} + c},$$

and not $\frac{a^2}{b^2 + c}$ at all. Our "simplification" has led to an inconvenient "three-storey" fraction, and of course, in this case, it should not be carried out.

Thus, it should be strictly remembered, that in simplifying algebraic fractions we may cross out only the same factors of the entire numerator and the entire denominator. If we do not observe this rule, we may easily arrive at a conclusion of the type:

$$\frac{2ab}{a+b} = \frac{2b}{1+b} = \frac{2}{1+1} = 1.$$

The error introduced in such a "simplification" leads to a violation of the distributive law (or distributive property) for the divisor in the division of an algebraic sum. This law states that in order to divide a sum by some number, it is necessary to divide by that number every term of that sum, and is expressed by the formula:

$$\frac{a+b}{c} = \frac{a}{c} + \frac{b}{c}.$$

Another widespread error consists in a term-by-term extraction of a root from a sum: it is considered that in order to extract the root from a sum, one should extract the root from every component separately, i.e. that $\sqrt{(a^2 + b^2)} = a + b$. Clearly this is untrue: $(a + b)^2$ is equal to $a^2 + 2ab + b^2$, and not to $a^2 + b^2$. The equality $\sqrt{(a^2 + b^2)} = a + b$ is an obvious absurdity also from the point of view of geometry—it expresses the equality of the hypotenuse to the sum of the legs in an arbitrary right triangle. It turns out that the operation of extraction of a root does not possess the distributive property with respect to addition and subtraction, but does possess it with respect to multiplication and division:

$$\sqrt{(a+b)} \neq \sqrt{(a)} + \sqrt{(b)}, \quad \sqrt{(ab)} = \sqrt{(a)} \times \sqrt{(b)},$$
$$\sqrt{(a:b)} = \sqrt{(a)} : \sqrt{(b)}.$$

We note that the operations of rank two (multiplication and division) possess the distributive property with respect to the results of the operations of rank one (addition and subtraction). The operations of rank three (raising to powers and extraction of the root) possess distributive properties with respect to the results of operations of rank two and do not possess them with respect to operations of rank one.

Every erroneous answer contradicts one of the original principles, or one of the earlier conclusions, of the given branch of knowledge. We may therefore formulate the question—Are there conditions which, if observed, will lead to the fact that assertions which are erroneous (in the general case) may turn out to be correct (in a specific case)?

In teaching practice some student's errors may be used as an excuse for carrying out very worthwhile (in the pedagogical sense) investigations. The students are to establish under what complementary conditions an erroneous relation will turn out to be correct. By such exercises they arrive at a better understanding of theory and of the error introduced. These exercises play a certain role in the development of the functional thinking of the student, since every expression is then considered as a function of the letters entering in it.

We shall make the above assertions more specific by analysing one example.

A student of grade VII, in carrying out the operation of the addition of two fractions, added their numerators and denominators separately, took the first sum as the numerator and the second one as the denominator, i.e. carried out the addition of fractions thus:

$$\frac{a}{b} + \frac{c}{d} = \frac{a + c}{b + d}. \qquad (1)$$

Ignoring for the moment the obvious incorrectness of formula (1), it is possible, if one understands the letters to mean arbitrary numbers, to find an infinitely large set of values for a, b, c and d for which the equality (1) does hold. To the number of such values belong, for example, the following: $a = -12$, $b = 2$,

$c = 3, d = 1$. Generally speaking, the equality (1) will be satisfied by any system of four numbers, taken under the condition that:

$$b \neq 0, \quad d \neq 0, \quad b + d \neq 0 \quad \text{and} \quad \frac{a}{c} = -\frac{b^2}{d^2}.$$

We arrive at the latter relation by means of the following elementary calculations:

$$\frac{a}{b} + \frac{c}{d} = \frac{a + c}{b + d}; \quad ad(b + d) + bc(b + d) = bd(a + c);$$

$$abd + ad^2 + b^2c + bcd = abd + bcd;$$

$$ad^2 + b^2c = 0; \quad ad^2 = -b^2c; \quad \frac{a}{c} = -\frac{b^2}{d^2},$$

or

$$a = -\frac{b^2c}{d^2}. \tag{2}$$

Giving arbitrary values to b, c and d, except $b = 0$, $d = 0$, $b + d = 0$, and finding the corresponding values of a by formula (2) we obtain an infinite set of systems of four numbers satisfying the formula (1).

The equality (1) is correct if, and only if, the additional condition (2) is satisfied. Such formulas, of course, have no practical significance. By means of them, however, one may exhibit a collection of mathematical curiosities.

The formula $\frac{a}{b} + \frac{c}{d} = \frac{ad + bc}{bd}$ is not subject to any restrictions, if we disregard the requirement that b and d are to be other than zero, expressing the ban on division by zero. Of course, this is the only formula expressing the rule for addition of algebraic fractions.

Now we make suggestions for similar investigations:

I. $$\frac{a + b}{ac} = \frac{1 + b}{c}.$$

In which special cases will this gross mistake not lead to confusion?

II. $$\frac{a}{b+c} = \frac{a}{b} + \frac{a}{c}.$$

The same question.

III. $$\sqrt{(a^2 + b)} = a\sqrt{(b)}.$$

An analogous question.

II. Analysis of the Examples

16. If we start from the fact that in mathematics, as a science, all operations are carried out on abstract numbers, then the passage from relation (1) to (2) does not make sense.

However, it is also possible to give another explanation of this error, by pointing out that there exist no "square dollars" and "square cents," and therefore the operation of extracting the square root of both members of relation (1) is meaningless.

The illusory plausibility of the argument is based on the ambiguity of the term "number"; in this case on the confusion of the concept of the abstract with that of a concrete number.

17. The student obviously forgot about the ambiguity of the term "root" (in the given case: actual and extraneous). In particular, he forgot that in solving an irrational equation extraneous roots may be introduced. This is why he considered the check only as a confirmation of the truth of his operations and calculations and not as an organic part of the required solution.

Without recognizing this, it is impossible to understand his astonishment when he obtained a false equality ($6 = 2$) which had seemed as if it might be correct.

We shall clarify the reason for the student's astonishment. With this aim we have to answer two questions:

(1) Why did the extraneous root appear?
(2) Of what equation is it a root?

To reach the greatest possible clarity, it is expedient to use a method of solution somewhat different from that used by the student. It consists in transforming equation (1) to the form $f(x) = 0$, and then multiplying the left- and the right-hand members by a multiplier which gives on the left-hand a difference of the squares of two functional expressions.

Thus, we represent equation (1) in the form

$$\sqrt{(x)} - (2 - x) = 0.$$

We multiply both members by the factor:

$$f_1(x) = \sqrt{(x)} + (2 - x).$$

Then, obviously, we obtain:

$$x - (2 - x)^2 = 0, \quad \text{or} \quad x^2 - 5x + 4 = 0.$$

We have reached the eqn (2), p. 63 of the original, whose roots are equal to 4 and 1.

The root $x = 4$ is extraneous for the eqn (1). It has appeared as a result of multiplying both members of the equation by the factor $f_1(x) = \sqrt{(x)} + (2 - x)$, whose root it is. As a matter of fact, $f_1(4) = \sqrt{(4)} + (2 - 4) = 2 - 2 = 0$.

The student's method of solution, of course, is equivalent to the method now considered, but, by means of the latter, the multiplier giving the extraneous roots is segregated more explicitly.

18. The absence of roots in eqn (1) is easily observed from the fact that its left-hand member $3\sqrt{(x)}$ is non-negative and is meaningful for $x > 0$, while the right-hand member $-x - 2$ is negative under those conditions. The solutions $x_1 = 4$ and $x_2 = 1$ are in fact roots of the function $f_1(x) = 3\sqrt{(x)} - (x + 2)$, as is easily seen by direct substitution.

19. A careful check of the proof of the divisibility of the difference $x^n - a^n$ by $a^{n-1}(x - a)$ reveals no error. In the same way, no doubts are called forth by the special cases considered. Yet for $x = 3$, $a = 2$, $n = 3$, the difference $x^n - a^n = 19$ is not divisible by $a^{n-1}(x - a) = 4$.

Let us try to find the remainder on dividing $x^n - a^n$ by $a^{n-1}(x - a) = a^{n-1}x - a^n$. It is equal to:

$$\frac{1}{a^{n-1}}x^{n-1} + \frac{1}{a^{n-2}}x^{n-2} + \frac{1}{a^{n-3}}x^{n-3} + \ldots + \frac{1}{a^2}x^2 + \frac{1}{a}x + 1.$$

The fractions appearing in all the terms of the remainder except the last put one on one's guard and immediately indicate the cause of the error: the phrase "is divisible" has different meanings in algebra and in arithmetic.

When in algebra we say that one polynomial in powers of some letter, for example x, "is divisible" by another polynomial in powers of the same x, what is meant is the possibility of obtaining a whole quotient, i.e. a polynomial also in powers of x. Whether the coefficients of the separate terms of the quotient are whole numbers or fractions is quite immaterial. For example, the binomial $ax^2 - a^2$ is divisible by $x - a$ and gives as the quotient $ax + a^2$. The coefficients a and a^2 may turn out to be either integers or fractions. The binomial $x^2 - a^2$ is divisible by $ax - a^2$ and yields as quotient $\frac{1x}{a} + 1$. The coefficient $\frac{1}{a}$ may even then have an integral value (for example with $a = \frac{1}{2}$) or a fractional value (for example, for $a = 2$).

In arithmetic, however, as applied to natural numbers, the words "is divisible" has a quite difference sense—if a natural number a is divisible by a natural number b, then there exists a natural number c, such that when multiplied by b it yields a.

Therefore from the divisibility of one polynomial by another (in the algebraic sense) still does not follow the divisibility (in the arithmetical sense) of those numbers which are obtained on substituting numbers for the letters in the polynomials. The divisibility of numbers follows from the divisibility of polynomials only when all the coefficients of the quotient are integral algebraic expressions and not fractions (with the condition, of course, that all the letters are replaced by natural numbers and that the divisor takes a value other than zero).

There is also another difference between the division of polynomials and the division of natural numbers. In the division of a polynomial by a polynomial the degree of the remainder (in the principal letter) is always lower than the degree of the divisor, whereas in division of natural numbers the remainder is always smaller than the divisor. The division of polynomials, carried out quite correctly may, upon replacement of the letters by numbers, lead to a case of division of numbers such as is not correct in the arithmetical sense. For example, on dividing the polynomial $x^3 - 3x^2 + 4x + 3$ by $x^2 - 2x + 1$, we obtain the quotient $x - 1$ and the remainder $x + 4$. If we substitute 3 for x, the dividend takes the value 15, the divisor 4, and the quotient and remainder are equal to $3 - 1 = 2$ and $3 + 4 = 7$, respectively; thus the remainder turns out to be greater than its divisor.

Thus, what is correct in the algebraic sense is not always correct from the point of view of arithmetic.

20. In the argument a serious error is introduced. From the equation $x - a = x$, where a is an arbitrary real number other than zero, it certainly does not follow that $a = 0$.

In fact, the solution of the equation $x - a = x$ leads to the following conclusions:

$$x - x = a, \quad \text{whence} \quad (1 - 1) \times x = a \quad \text{or} \quad 0 \times x = a.$$

Since, according to the data, $a \neq 0$, then the equation $0 \times x$ does not have solutions: there exists no number which when multiplied by zero yields a number other than zero.

However, it is possible that the reader has felt some doubt: is the transition from $(x - a)^3 = x^3$ to $x - a = x$ lawful?

Such a transition is entirely lawful, if the numbers cubed are both real: the cube root of a real number, for real numbers, has only one value (positive, if the number under the radical sign is positive, and negative if it is negative).

From the above it follows that the result obtained—eqn (2) has no roots—indicates the absence of roots within the set of real numbers for eqn (1).

21. The root of eqn (1) is the number 10, as we may easily convince ourselves.

For $x = 10$, the relation (2) takes the form:

$$\frac{0}{-3} = \frac{0}{3}$$

Since the quotient on dividing zero by any number, other than zero, is equal to 0, then, obviously, from the relation $a/b = a/d$ one may not deduce the equality of b and d, if $a = 0$.

Here we meet the use of a false analogy, namely the extension of some proposition (if $a/b = a/d$ and $a \neq 0$, then $b = d$) to an exceptional case ($a = 0$).

22. In the set of real numbers the relation (1) has no meaning: every power of a positive number is a positive number.

Relation (1) takes on a meaning, if we consider the problem in the set of complex numbers. In that case, setting $b = i$ and $a = 2$ we have the correct relation $i^2 = -1$, which, as we know, does not lead to contradictions.

However, with such a formulation of the problem we go beyond the limits within which the erroneous argument was being carried out.

23. The concept of an arithmetic root is not introduced in the set of complex numbers.

The correct calculation of the product $\sqrt{(-1)} \times \sqrt{(-1)}$ assumes the following course of reasoning:

$$\sqrt{(-1)} \times \sqrt{(-1)} = (\pm i) \times (\pm i) = \pm i^2 = \pm 1,$$

which leads to the same result as the extraction of a square root from the unity: $\sqrt{(1)} = \pm 1$.

Now if we take for both factors of the product $\sqrt{(-1)} \times \sqrt{(-1)}$ the same value of the square root of minus one, then we obtain -1 as result.

In fact:

$$\sqrt{(-1)} \times \sqrt{(-1)} = i \times i = i^2 = -1;$$

$$\sqrt{(-1)} \times \sqrt{(-1)} = (-i) \times (-i) = (-i)^2 = -1.$$

24. Both in this argument and in the argument of problem 23, we have allowed ourselves to forget the fact that in the set of complex numbers the concept of the arithmetic root is not introduced.

The transformations carried out in line (2) are erroneous.

For a thorough study of the mistakes introduced here we shall contrast the true argument with the false one:

$$x = \sqrt[4]{((-1) \times (-1))} = \sqrt{(\sqrt{(-1)} \times (-1))}$$
$$= \sqrt{(\sqrt{(-1)} \times \sqrt{(-1)})} = \sqrt{((\pm 1) \times (\pm 1))}.$$

If the factors of the expression under the radical are taken with the same signs, then $x = \sqrt{(i^2)} = \pm i$.

If the factors of the expression under the radical are taken with differing signs then $x = \sqrt{(1)} = \pm 1$.

Consequently, as a result of extracting the fourth root of the product $(-1) \times (-1)$ four values of x are obtained, namely 1, -1, i, $-i$.

It is easy to check the correctness of the results. In fact, the numbers 1, -1, i, and $-i$ should satisfy the equation $x^4 = 1$, and this is indeed true: $1^4 = (-1)^4 = i^4 = (-i)^4 = 1$.

The same result may be obtained by solving the equation $x^4 - 1 = 0$ by the method of factoring its left-hand number:

$$(x - 1) \times (x + 1) \times (x^2 + 1) = 0.$$

Thus, the eqn (1) is satisfied by each of the four numbers: 1, -1, i, $-i$. One of the values of x, naturally, should agree with the original assumption, which does, indeed, hold.

The untrue conclusion is rejected.

We shall now make the problem somewhat broader, considering the rule for the multiplication of square roots not only of negative numbers, but also of complex numbers, i.e. of numbers of the form $z = x + iy$, where x and y are real numbers. As is known from the mathematics course of the tenth grade, every such number is represented by a definite point of a plane, namely the point with the abscissa x and ordinate y, and may be represented in the trigonometric form $z = r(\cos\alpha + i\sin\alpha)$, where the "modulus" r is a positive (or zero) real number expressing the distance of that point from the origin of coordinates, and the "argument" α expresses the angle between the positive direction of the x-axis and a ray drawn from the origin to that point (Fig. 14). The angle α, taken in degrees or in radians, may be expressed by any real number (positive, negative, or zero), but this number may be always "reduced" to the interval 0–360° by adding or subtracting a whole multiple of 360°, and therefore we shall assume that $0° \leqslant \alpha < 360°$.

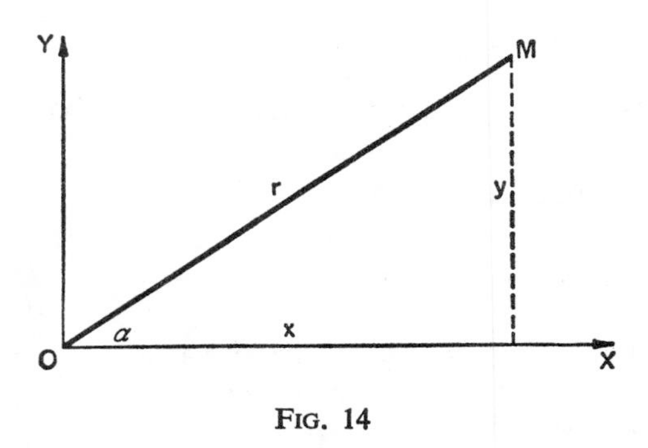

FIG. 14

The multiplication law for complex numbers, expressed in trigonometric form, is given by the well-known De Moivre formula:

if $\quad z = r(\cos\alpha + i\sin\alpha), \quad z_1 = r_1(\cos\alpha_1 + i\sin\alpha_1),$

then $\quad zz_1 = rr_1[\cos(\alpha + \alpha_1) + i\sin(\alpha + \alpha_1)],$

which is easily obtained on multiplying the expressions for z and for z_1, applying the usual rule for multiplication of polynomials, replacing i^2 by -1, and then making use of the formulas for the sine and the cosine of the sum of two angles.

We now find the square root of the number

$$z = r(\cos \alpha + i \sin \alpha).$$

Assuming that it is equal to some complex number

$$\omega = \rho(\cos \phi + i \sin \phi),$$

we may find the unknowns ρ and ϕ from the given r and α. From the condition

$$\omega = \sqrt{(z)}, \quad \text{whence} \quad \omega^2 = z.$$

According to the De Moivre formula:

$$\omega^2 = \rho\rho[\cos (\phi + \phi) + i \sin (\phi + \phi] = \rho^2(\cos 2\phi + i \sin 2\phi).$$

We have the equality:

$$\rho^2(\cos 2\phi + i \sin 2\phi) = r(\cos \alpha + i \sin \alpha).$$

Two complex numbers given in the trigonometric form are equal (and represent one and the same point of the plane) if, and only if, their moduli are equal, and the arguments are either equal, or differ from each other by a whole multiple of 360°. Hence we conclude that $\rho^2 = r$, $2\phi = \alpha + 360° \times n$, where n is an arbitrary integer. The modulus ρ, as a real non-negative number, is equal to the arithmetic (positive) value of the square root of the non-negative real number r : $\rho = +\sqrt{(r)}$; and the argument ϕ has an infinite set of values, defined by the formula $\phi = \frac{1}{2}\alpha + 180° \times n$ for an arbitrary integral n. However if from this whole infinite set of values of ϕ we omit those which differ from the others by a whole multiple of 360°, then there remain only two values, corresponding to the numbers $n = 0$ and $n = 1$, namely:

$$\phi_1 = \tfrac{1}{2}\alpha, \quad \phi_2 = \tfrac{1}{2}\alpha + 180°,$$

and we arrive at the conclusion that there exist two and only two values of the square root of the number

$$z = r(\cos\alpha + i\sin\alpha),$$

namely:

$$\omega_1 = +\sqrt{(r)}\,(\cos\tfrac{1}{2}\alpha + i\sin\tfrac{1}{2}\alpha),$$

$$\omega_2 = +\sqrt{(r)}\,[\cos(\tfrac{1}{2}\alpha + 180^\circ) + i\sin(\tfrac{1}{2}\alpha + 180^\circ)]$$

$$= -\sqrt{(r)}\,(\cos\tfrac{1}{2}\alpha + i\sin\tfrac{1}{2}\alpha),$$

where

$$\omega_2 = -\omega_1.$$

In Fig. 15 the point A represents the given number z, and the points B_1, B_2 represent both values of $\sqrt{(z)}$ (here $OA = r$, $OB_1 = OB_2 = +\sqrt{(r)}$); the point B_1 is the first value of $\sqrt{(z)} = \omega_2 = -\omega_1$.

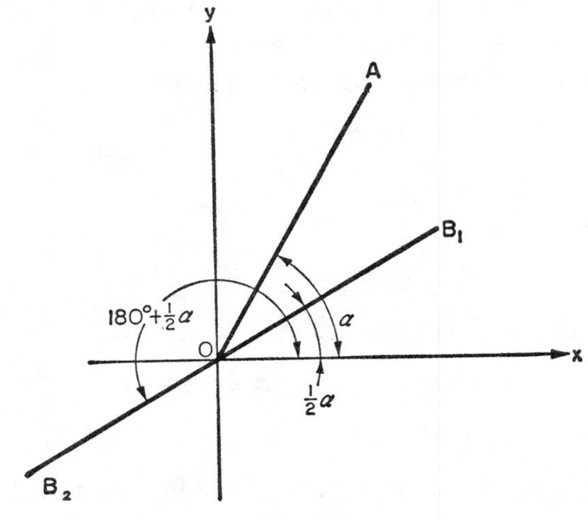

FIG. 15

The argument α, as we have seen above, may be always considered as satisfying the inequality $0^\circ \leqslant \alpha < 360^\circ$, and therefore

$$0^\circ \leqslant \tfrac{1}{2}\alpha < 180^\circ,\ 180^\circ \leqslant \tfrac{1}{2}\alpha + 180^\circ < 360^\circ.$$

Thus, of the two values of the square root, one is always represented by a point of the upper half-plane (i.e. of that portion of the plane which is situated above the axis of abscissas) or in the limiting case (for $\alpha = 0°$), by a point of the positive portion of the x-axis, and the other—always by a point of the lower half-plane, or in the limiting case (for $\alpha = 0°$), by a point of the negative portion of the x-axis.

In the case when the root is extracted from a real positive number, we have $\alpha = 0°$, and the first value of ω_1 is simply the arithmetic value of the root $+\sqrt{(z)}$. Therefore, one of the two values of the square root of the complex number z, namely that value of it which is expressed by the point of the positive x-semiaxis, or by a point of the upper half-plane, we shall call "the first" or "arithmetical" value of the root. The other value of the root, namely that which is represented by a point of the negative x-semiaxis, or of the lower half-plane, we shall call "the second" or "non-arithmetic" value of the root.

Take now two arbitrary complex numbers:

$$z = r(\cos \alpha + i \sin \alpha)$$

and

$$z' = r'(\cos \alpha' + i \sin \alpha'),$$

where

$$0° \leqslant \alpha < 360°, \quad 0° \leqslant \alpha' < 360°,$$

and find the "first" values of their square roots:

$$\omega_1 = +\sqrt{(r)}\,(\cos \tfrac{1}{2}\alpha + i \sin \tfrac{1}{2}\alpha),$$

$$\omega'_1 = +\sqrt{(r')}\,(\cos \tfrac{1}{2}\alpha' + i \sin \tfrac{1}{2}\alpha').$$

Multiply them together, applying De Moivre's rule:

$$\omega_1\omega'_1 = (+\sqrt{(r)})(+\sqrt{(r')})\left(\cos \frac{\alpha + \alpha'}{2} + i \sin \frac{\alpha + \alpha'}{2}\right)$$

$$= +\sqrt{(rr')}\left(\cos \frac{\alpha + \alpha'}{2} + i \sin \frac{\alpha + \alpha'}{2}\right).$$

Further, we find the product of the numbers z and z', and both values of its square root:

$$zz' = rr'[\cos(\alpha + \alpha') + i \sin(\alpha + \alpha')],$$

$$u = \sqrt{(zz')} = +\sqrt{(rr')}\left[\cos\frac{\alpha + \alpha'}{2} + i \sin\frac{\alpha + \alpha'}{2}\right],$$

$$v = \sqrt{(zz')} = +\sqrt{(rr')}\left[\cos\left(\frac{\alpha + \alpha'}{2} + 180^\circ\right)\right.$$

$$\left. + i \sin\left(\frac{\alpha + \alpha'}{2} + 180^\circ\right)\right].$$

Now we have everything ready for the solution of the problem, whether the product of first values of the square root of two arbitrary complex numbers gives the "first" value of the square root from the product of these numbers, or not.

As we see, the product of the "first" values of the square root of the numbers

$$z = r(\cos\alpha + i \sin\alpha) \text{ and } z' = r'(\cos\alpha' + i \sin\alpha'),$$

and namely

$$\omega_1\omega'_1 = +\sqrt{(rr')}\left(\cos\frac{\alpha + \alpha'}{2} + i \sin\frac{\alpha + \alpha'}{2}\right),$$

is equal to one of the values of the square root of the number zz', namely to the number:

$$u = +\sqrt{(rr')}\left(\cos\frac{\alpha + \alpha'}{2} + i \sin\frac{\alpha + \alpha'}{2}\right).$$

Now the question arises whether this latter number is the "first" value of the root of zz', or whether this "first" value is the second value of the root of zz', designated by us by the letter v.

From the conditions $0^\circ \leqslant \alpha < 360^\circ$ and $0^\circ \leqslant \alpha' < 360^\circ$, we conclude that $0^\circ \leqslant \alpha + \alpha' < 720^\circ$ and $0^\circ \leqslant \frac{1}{2}(\alpha + \alpha') < 360^\circ$.

In other words, the point representing the number u may lie either on the positive portion of the x-axis $\left(\text{with } \frac{\alpha + \alpha'}{2} = 0^\circ\right)$ or above the x-axis $\left(\text{with } 0^\circ < \frac{\alpha + \alpha'}{2} < 180^\circ\right)$, or on the negative portion of the x-axis $\left(\text{for } \frac{\alpha + \alpha'}{2} = 180^\circ\right)$, or below the x-axis $\left(\text{for } 180^\circ < \frac{\alpha + \alpha'}{2} < 360^\circ\right)$. Thus the product of the "first" values of the roots of z and z' may be equal in some cases to the "first" value of the root of zz', and in other cases to its "second" value.

By considering instead of the values of half the sum of the arguments $\frac{1}{2}(\alpha + \alpha')$, the value of the whole sum $\alpha + \alpha'$, and collecting in pairs the four cases considered above, we arrive at the following two cases:

Case I. If $\alpha + \alpha' < 360^\circ$, then either $\frac{\alpha + \alpha'}{2} = 0^\circ$, or $0^\circ < \frac{\alpha + \alpha'}{2} < 180^\circ$; the number u is represented either by a point of the positive portion of the x-axis, or by a point of the upper half-plane; u is equal to the "first" value of the root of zz'.

Case II. If $\alpha + \alpha' > 360^\circ$, then either $\frac{\alpha + \alpha'}{2} = 180^\circ$, or $180^\circ < \frac{\alpha + \alpha'}{2} < 360^\circ$, the number u is represented either by a point of the negative portion of the x-axis, or by a point of the lower half-plane; u is equal to the "second" value of the root of zz'.

We now obtain the final conclusion: the product of the "first" values of square roots of complex numbers z and z' is equal to the "first" value of the square root of the product of these numbers if, and only if, the sum of the arguments of these numbers is less than 360° (it is assumed that the arguments are selected within the limits 0–360°).

Noting that the "second" value of the square root is equal to its "first" value, taken with the opposite sign, we can, by making use of the last proposition, easily construct the following "table of multiplication" for square roots of complex numbers (the "first" values of the square root we shall denote by the plus sign in front of the square root symbol, and the "second" value by the minus sign). Just as before, we assume that the arguments α and α' of the numbers z and z' are taken between 0° and 360°.

Case I, $\alpha + \alpha' < 360°$

$$(+\sqrt{(z)}) \times (+\sqrt{(z')}) = +\sqrt{(zz')},$$
$$(+\sqrt{(z)}) \times (-\sqrt{(z')}) = -\sqrt{(zz')},$$
$$(-\sqrt{(z)}) \times (+\sqrt{(z')}) = -\sqrt{(zz')},$$
$$(-\sqrt{(z)}) \times (-\sqrt{(z')}) = +\sqrt{(zz')}.$$

Case II, $\alpha + \alpha' > 360°$

$$(+\sqrt{(z)}) \times (+\sqrt{(z')}) = -\sqrt{(zz')},$$
$$(+\sqrt{(z)}) \times (-\sqrt{(z')}) = +\sqrt{(zz')},$$
$$(-\sqrt{(z)}) \times (+\sqrt{(z')}) = +\sqrt{(zz')},$$
$$(-\sqrt{(z)}) \times (-\sqrt{(z')}) = -\sqrt{(zz')}.$$

We now return to the special case of the multiplication of roots with which we started: let z and z' be two real negative numbers. Every such number may be written in the form $z = x + i \times 0$, where x is a real negative number, and therefore such a number z is represented by a point of the negative semiaxis x. Consequently,

$$\alpha = 180°, \quad \alpha' = 180°, \quad \alpha + \alpha' = 360°,$$

and here we have case II: the product of the "first" values of the square roots of two negative real numbers is equal to the "second" value of the square root of the product of these numbers.

25. To find the error, we contrast the correct solution with the incorrect argument that led to the absurd conclusion.

Proceeding with the relation $\frac{1}{-1} = \frac{-1}{1}$ and applying the operation of extracting the square root, we have:

$$\frac{\sqrt{(1)}}{\sqrt{(-1)}} = \frac{\sqrt{(-1)}}{\sqrt{(1)}},$$

i.e.

$$\frac{\pm 1}{\pm i} = \frac{\pm i}{\pm 1},$$

whence

$$(\pm 1) \times (\pm 1) = (\pm i) \times (\pm i).$$

With the proper choice of signs, this relation does not lead to any contradiction. In fact:

$$1 \times 1 = (-1) \times (-1) = i \times (-i) = (-i) \times i = -i^2 = 1;$$

$$1 \times (-1) = (-1) \times 1 = i \times i = (-i) \times (-i) = i^2 = -1.$$

26. If we examine carefully the whole argument for the error leading to such an absurd conclusion, we discover that it is contained in that "obvious fact" with whose formulation we began. For positive numbers, that assertion is correct—if $a : b = c : d$, and $a > b$, where all the four numbers a, b, c, and d are positive, then, setting $a : b = q$, we conclude, that $q > 1$. However, the ratio $c : d$ is also equal to q, and if the ratio of two positive numbers is greater than 1, then the first number is greater than the second, and consequently $c > d$.

For rational numbers, as the present sophism shows, this conclusion may not be true. In addition to the proportion considered, $(+a) : (-a) = (-a) : (+a)$, we may give any number of others, for which the "obvious fact" of the beginning of our section is not confirmed. For example:

$$15 : 3 = (-15) : (-3),$$

$$15 : (-3) = (-100) : 20.$$

In the present sophism we have a very instructive example of how easy it is to err by trusting an "obvious fact," i.e. by accepting without proof some assertion that seems correct at first sight. In taking some assertion without proof, we have to enter it in the list of axioms. All the assertions not included in the list of axioms should be regarded as theorems and be subject to proof (on the basis of the axioms adopted and the definitions, and also of theorems proved earlier). It would be good to eliminate the word "obvious" from use in mathematical arguments altogether; instead of referring to the "obvious," we ought really to refer to a definite axiom, or a proof, i.e. the reduction of the new assertion to the axioms and theorems proved earlier.

On the other hand, the present sohpism shows an example of how a proposition proved for some numbers (in this case, positive numbers) may cease being true for numbers of a more general type (in this case, rational numbers). In passing from integers to fractions, the student, having become accustomed to the fact that multiplication by an integer is connected with an increase (for a multiplier greater than 1) and division with a decrease of the number taken, only with difficulty becomes accustomed to the fact that in multiplication and division by proper fractions everything occurs the other way around.

In what follows we shall meet a number of examples of errors occurring from a similar application of theorems, correct for some conditions, to cases where those conditions are not fulfilled.

27. It is easy to convince oneself, even by taking numerical values, that the error is made on passing from the inequality (1) to the inequality (2), i.e. on dividing both members of the inequality (1) by the difference $b - a$, which for $a > b$ has a negative value. The point is that division of both members of the inequality by one and the same number leads to an inequality in the same direction, i.e. to the inequality with the same one of the two symbols ($>$ and $<$), only if the divisor is positive. For a negative divisor the sense of the inequality changes. The proof of this property of inequalities may be found in any textbook of algebra. If in passing from the inequality (1) to the inequality (2), we take into consideration the change of the sense of the inequality, we obtain that $a < b + a$, and the untrue conclusion that $a > 2b$ is rejected.

28. The theorem on the multiplication of inequalities, given at the beginning of Sec. 28, is formulated inexactly: it is true only for inequalities in which all the terms are positive. Here is its exact formulation: one may multiply, term by term, two inequalities of the same sense, if all their members are positive; the new inequality will have the same sense as each of the given inequalities. If this theorem is applied to inequalities such as $5 > -1$ and $2 > -15$, it is possible to obtain the absurdity: $5 \times 2 > (-1) \times (-15)$, i.e. $10 > 15$. The careless multiplication of inequalities has led to the absurd conclusion, that $a > b$ and $b > a$.

29. The error is introduced in the multiplication of both members of the inequality $a > b$ by $b - a$. The point is that the difference of the numbers b and a is negative, since, according to the condition, $a > b$. Consequently, in multiplying by $b - a$, i.e. by a negative number, one should change the sense of the inequality. However, this has not been carried out, and has led to the absurd conclusion.

30. The error is introduced in passing from the equality (2) to the inequality (3). Here one had clearly forgotten the fact that, since the logarithm of a proper fraction is negative, one should set in relation (3) not the symbol for greater than, but, inversely, the symbol for less than—the doubled negative number is less than that negative number itself.

31. By carefully checking both solutions, we find only one doubtful point: are we right in considering that $+\sqrt{(1-x)^2} = 1 - x$? In other words: is the arithmetic value of the square root of the square of some number always equal to that number, is $+\sqrt{(a^2)} = +a$ always? The last formula, is, obviously, true for $a \geqslant 0$, but for $a < 0$ it should be replaced by another, namely by the formula $+\sqrt{(a^2)} = -a$, since, if a is negative, then to obtain the arithmetic (positive) value of the root, it is necessary to take not a, but $-a$. Both of these formulas may be combined in the one:

$$+\sqrt{(a^2)} = |a|.$$

Solving the problem posed above by the second method, we have taken $+\sqrt{(1-x)^2} = 1 - x$. This is true if $1 - x \geqslant 0$, i.e. if $x \leqslant 1$. But in what followed we were replacing x by the number 3, i.e. were making the difference $1 - x$ negative. This means that, in order to obtain a positive value of the square root of $(1 - x)^2$ for such a value of x, we should take not $1 - x$ but $-(1 - x)$ or $x - 1$. The following solution will be correct:

$$2x + \sqrt{(1-x)^2} = 2x + [-(1-x)] = 2x - 1 + x$$
$$= 3x - 1 = 3 \times 3 - 1 = 8.$$

We have obtained the same thing as in the calculation by the first method.

32. The equality $\log x^2 = 2 \log x$ is true if $x > 0$, and untrue if $x < 0$.

If $x < 0$, then $\log x^2 = 2 \log(-x)$.

In general, if $x \neq 0$, then there holds the equality:

$$\log x^2 = 2 \log |x|.$$

From the above it is clear that in the argument under consideration, from the relation $a^2 = b^2$ (2), there should follow not the relation $2 \log a = 2 \log b$ (3), which is false, but the relation:

$$2 \log |a| = 2 \log |b|.$$

From the last equality we conclude:

$$|a| = |b|,$$

which does not contradict the relation $a = -b$ (1), and gives no basis for absurd conclusions.

The false reasoning under analysis has played such an important part in the history of the development of mathematics that one should dwell on this problem in somewhat greater detail.

In the history of mathematics of the 18th century a considerable place, from the point of view of its scientific and methodological interest, is taken by the discussion of whether there exist real values of logarithms of negative numbers.

In elaborating the problem of integration, Leibnitz (1646–1716) and Johann Bernoulli (1667–1748) encountered the problem of logarithms of negative numbers. At first, concerned only with developing definite algorithms for integration, they approached the taking of logarithms of numbers of a given nature in a purely formal fashion. But in 1712–1713 there arose between them a lively dispute in the form of correspondence concerned with the very essence of the problem. In his letters Bernoulli disputes Leibnitz's opinion that logarithms of negative numbers are imaginary, asserting that they are real, since, according to his conviction, $\log x = \log(-x)$.

The correspondence between Leibnitz and Bernoulli was published in 1745. Upon familiarizing himself with it, Leonard Euler (1707–1783) published in 1749 an article "On the Dispute between Bernoulli and Leibnitz as to the Logarithms of Negative and Imaginary Numbers," in which he gave a brilliant solution of the problem. Having established that Leibnitz was right, Euler was not satisfied with the latter's argument, arriving at the solution of the problem by means of his formula for imaginary exponents.

However, the publication of Euler's work still did not end the dispute. An authoritative support for Bernoulli's opinion was given by D'Alembert (1717–1783). He published it in 1761 in a polemic paper "On the logarithms of Negative Numbers" and in a paper written, in the same spirit, on logarithms for volume XX of the famous *Encyclopedia* (1778). The point of view of Bernoulli–D'Alembert was shared also by some other mathematicians of the 18th century.*

* The history of this problem is well set forth in considerable detail in the book of A. I. Markushevich, *Elements of the Theory of Analytic Functions* Elementy teorii analiticheskikh funktsiy), pp. 42–46, Moscow, 1944.

An echo of this interesting dispute is a very widespread sophism retained to this day, bearing the name of Bernoulli.

Bernoulli, proceeding from the indisputable equality $(-a)^2 = (+a)^2$, asserted, that since $\log(-a)^2 = \log(+a)^2$, then

$$2\log(-a) = 2\log(+a),$$

and therefore $\log(-a) = \log(+a)$. Hence he drew the conclusion that the logarithms of negative numbers have real values.

Here in passing from the equality $\log(-a)^2 = \log(+a)^2$ to the equality $2\log(-a) = 2\log(+a)$, an error is introduced. The point is that, while the sets of values of $\log(-a)^2$ and $\log(+a)^2$ coincide, the sets of values of $2\log(-a)$ and $2\log(+a)$ not only do not coincide, but even have no common subset.

We analyse this problem in somewhat greater detail:

From Euler's formula for imaginary exponents

$$e^{iy} = \cos y + i\sin y,$$

we note directly, that $e^{2n\pi i} = 1$ and $\log 1 = 2n\pi i$, where n equals $0, \pm 1, \pm 2, \ldots$

From the formula giving the trigonometric form of the complex number $z = r(\cos\phi + i\sin\phi) = re^{i\phi} = re^{i(\phi+2n\pi)}$, where n is equal to $0, \pm 1, \pm 2, \ldots$, we find that $\log z = \log r + i(\phi + 2n\pi)$, whence $\log(-1) = \log 1 + i(\pi + 2n\pi) = i(2n+1)\pi$.

The possibility of the equality $\log(+1)$ and $\log(-1)$ is excluded, since the sets $\{2n\pi i\}$ and $\{(2n+1)\pi i\}$, where n is equal to $0, \pm 1, \pm 2, \ldots$, do not intersect, i.e. do not have even a single common element.

Thus, from the equality $\log(-1)^2 = \log(+1)^2$, it is invalid to conclude that $2\log(-1) = 2\log(+1)$. From the equality we can conclude only that $\log(-1) + \log(-1) = \log(+1) + \log(+1)$. Here in each member there occur as terms, in general, unequal values of $\log(-1)$ and unequal values of $\log(+1)$. In fact, passing from the last relation to the relation:

$$i(2n_1 + 1)\pi + i(2n_2 + 1)\pi = 2n_3\pi i + 2n_4\pi i,$$

we see that, for example, for $n_1 = 0$, $n_2 = 0$, $n_3 = 0$, $n_4 = 0$ the equality is satisfied.

33. The error of the present false proof is easily discovered by investigating the differences $a - a_1$, $b - b_1$, $c - c_1$, which arise on subtracting N_1 from N. Will these differences be numbers in the ordinary meaning of that word, i.e. is each of them one of the numbers 0, 1, 2, 3, 4, 5, 6, 7, 8, 9? If $N = 257$, $N_1 = 725$, then

$$a = 2,$$
$$b = 5,$$
$$c = 7,$$
$$a_1 = 7,$$
$$b_1 = 2,$$
$$c_1 = 5,$$
$$a - a_1 = -5,$$
$$b - b_1 = 3,$$
$$c - c_1 = 2,$$

and the difference $a - a_1$ is negative. It is easy to see that, if the numbers N and N_1 have one and the same sum of digits, then at least one of the differences $a - a_1$, $b - b_1$, $c - c_1$ is negative (an exception is formed only by the case when $a = b = c = a_1 = b_1 = c_1$ and when the transposition of the digits does not in fact change the number). As a matter of fact, rewriting the equality $a + b + c = a_1 + b_1 + c_1$ in the form

$$(a - a_1) + (b - b_1) + (c - c_1) = 0,$$

we find that, if one of these differences is positive, then at least one of the two others must have a negative value.

Thus, writing the difference $N - N_1$ in the form $100(a - a_1) + 10(b - b_1) + (c - c_1)$, we have no right to assert that in the number $N - N_1$ there are $a - a_1$ hundreds, $b - b_1$ tens, and

$c - c_1$ units. Instead of the differences we should take their absolute values:

$$|a - a_1|,$$
$$|b - b_1|,$$
$$|c - c_1|.$$

The sum of the digits of the number $N - N_1$ is equal not to the algebraic sum

$$(a - a_1) + (b - b_1) + (c - c_1),$$

but to the arithmetic sum:

$$|a - a_1| + |b - b_1| + |c - c_1|,$$

and all the subsequent transformations carried out in our "proof" are invalidated.

34. The misunderstanding is very simply solved. The theorem on the existence of a root in the algebra of complex numbers refers only to an equation of the form

$$a_0x^n + a_1x^{n-1} + a_2x^{n-2} + \ldots + a_{n-1}x + a_n = 0, \qquad (2)$$

where n is some natural number, i.e. has one of the values 1, 2, 3, 4, . . ., and all the coefficients $a_0, a_1, a_2, \ldots, a_{n-1}, a_n$ are arbitrary complex numbers, where $a_0 \neq 0$. Regarding such an equation, it has been proved that it always has at least one root (real or imaginary). It should be strictly recalled, that this theorem of higher algebra (or the theorem of Gauss, as it is called after the name of the great German mathematician who almost indisputably was the first to demonstrate it) refers only to equations of the form (2) and that it says nothing about equations of another form such, as for example, eqn (1). The stipulation made above about the value of the coefficients is also to the point. If, for example, we take the equation

$$0 \times x + 1 = 0, \qquad (3)$$

then it will turn out that it has no roots. Whatever number we take for x, the product $0 \times x$ will always be equal to zero, and the left-hand member of the equation will always turn out to be equal to 1. The equality of the left- and the right-hand members is thus impossible for any value of x whatever. This state of affairs in no way contradicts the Gauss theorem. In eqn (3) the coefficient of the leading term a_0 is 0, while in the conditions of the theorem it is indicated that $a_0 \neq 0$.

From the Gauss theorem it is easy to derive the corollary, stating that the left-hand member of eqn (2) may always be expanded into n linear factors, i.e. that the eqn (2) may always be represented in the form

$$a_0(x - \alpha_1)(x - \alpha_2)(x - \alpha_3) \ldots . (x - \alpha_n) = 0. \qquad (4)$$

The numbers $\alpha_1, \alpha_2, \alpha_3, \ldots, \alpha_n$, which may be either real (positive, equal to zero, or negative), or complex, either unequal or equal to each other, represent precisely the roots of equations (2) and (4). The substitution for x in eqn (4) of one of these numbers $\alpha_1, \alpha_2, \alpha_3, \ldots, \alpha_n$ turns eqn (4) into an identity $0 = 0$. Therefore this corollary of the Gauss theorem may be formulated otherwise—every equation of the form (2) has (for $a_0 \neq 0$) exactly n roots, i.e. as many roots as there are unities in the exponent of its degree.

Knowing the corollary of the Gauss theorem only in its last formulation, one may be perplexed on meeting, for example, the equation $x^3 = 0$. It has only one root $x = 0$, since no number not equal to zero, neither real nor imaginary, will yield zero when cubed. The corollary to the Gauss theorem in its first formulation, as applied to the equation $x^3 = 0$, states, that the left-hand member of that equation may be expanded into three linear factors of the form $x - \alpha$. But this is very easy to do: $x^3 = (x - 0)(x - 0)(x - 0)$. Now it becomes clear that the equation $x^3 = 0$ has three roots, but that all these roots are equal to each other.

When applying the corollary of the Gauss theorem, we should never forget the possibility of the equality of some (or even all) the roots of the equation.

35. In all the examples that we considered, both in the first two, when the method gave the correct results, and in the latter two, where it led to untrue conclusions, all the transformations called forth no doubts: everything was done in strict compliance with the rules of algebra. The only operation for which we cannot exhibit a basis in the form of some rule of algebra is the designation of the result of the infinite set of operations by the letter x, upon which we subsequently carry out operations, assuming, of course, that that letter x designates some definite, though, so far, unknown result. See, all the rules of algebra refer to operations upon numbers! The reason for the appearance of absurd conclusions in the results of our arguments in the third and fourth examples, lies in the fact that in those examples there does not exist a definite number which would be obtained after an infinite repetition of the operations under consideration. Expressing this more precisely, we may say that in the third example we are dealing with an infinite sequence of numbers

$$x_1 = a,$$
$$x_2 = a + a = 2a,$$
$$x_3 = a + a + a = 3a,$$
$$x_4 = a + a + a + a = 4a, \ldots,$$

which for $a > 0$ has no limit, since these numbers increase without bound—however great a given number A, we may always find a number n such that all the terms of the sequence, for numbers greater than n, will be greater than A (for this it is sufficient to take $n \geqslant \frac{A}{a}$). Denoting the "sum" of the infinite set of equal numbers by the letter x and carrying out operations upon x as if it denoted a number, we arrive at an absurd conclusion.

The fourth example is similar. Here we have a sequence of numbers $x_1 = 1$, $x_2 = -1$, $x_3 = 3$, $x_4 = -5$, $x_5 = 11$, . . ., which has no limiting value. The very designation of that non-existent limiting value by the letter x was incorrect, as well as the carrying out of operations upon it as upon a definite number.

The question now arises: how does one recognize whether a limiting value exists in an infinite sequence under consideration? The answer to that question is given by various tests for the existence of a limit of infinite sequences. A simple indication of this kind is studied in the ninth grade: if there is a given infinite sequence of increasing numbers

$$a_1,\ a_2 > a_1,\ a_3 > a_2,\ a_4 > a_3,\ a_5 > a_4, \ldots,$$

but each of these numbers is less than some number b, then the numbers $a_1, a_2, a_3, \ldots$ tend to a limit which is either less than b or equal to b.

In the first example considered above we had a sequence of an infinite set of increasing numbers:

$$x_1 = 0{\cdot}24;\ x_2 = 0{\cdot}2424;\ x_3 = 0{\cdot}242424;\ \ldots,$$

but each of these numbers is less than, for example, 0·3. Consequently, the limiting value exists. Denoting it by the letter x, we may securely operate upon x as was done above, and we arrive at the conclusion that $x = \frac{8}{33}$. In exactly the same way also in the second example we have an infinite sequence of increasing numbers:

$$x_1 = \sqrt{(2)},\quad x_2 = \sqrt{(2 + x_1)},\quad x_3 = \sqrt{(2 + x_2)},$$
$$x_4 = \sqrt{(2 + x_3)}, \ldots,$$

but each of these numbers is less than 2, since every expression under the square root sign is less than 4. Consequently here also the limiting value exists, and it may be denoted by x.

Thus, in applying the method of "eliminating infinity," we may feel sure of the correctness of the result only when we first establish that the limiting expression does indeed exist.

36. The reader who has already familiarized himself with Problem 35 of course has long since understood what is the matter: our sum of the infinitely great set of terms has no definite value whatever, since the successive summing of an even greater number of terms does not give an approximation to any limiting value:

$$x_1 = 1, \quad x_2 = 1 - 1 = 0, \quad x_3 = 1 - 1 + 1 = +1,$$
$$x_4 = 1 - 1 + 1 - 1 = 0, \ldots$$

The error of the first argument which led to the value 0·5 lies in the fact that we carried out operations upon x without having first established that the letter x denoted any definite number.

In the second argument, which gave $x = 0$, we made use of the associative property of algebraic addition: in any algebraic sum consisting of a definite number of terms we may include an arbitrary number of terms in brackets. The carrying over of this associative property to the sum of an infinite set of terms is not justified in any way and the result obtained (the definite numerical value, namely 0, for the expression which clearly has no definite numerical value), shows that such a carry-over may lead to an untrue conclusion.

In the third argument, which led to the value $x = 1$, we again made unlawful use of that associative property of algebraic addition as applied to the sum of an infinite set of terms.

In the fourth argument, in which we found that $x = -1$, we have again the same unlawful application of the associative property, but, besides, an unlawful application of the commutative property: as is well known, the value of an algebraic sum of a definite number of terms is not changed by a commutation of the terms, provided every term is taken with the same sign which stands in front of it. By transposing the components of a sum containing an infinite number of terms, in some cases we change the value of that sum.

Thus, all four of our arguments constitute a compact pile of errors based on the application to the sum of an infinite set of components, of methods that are lawful only when applied to the sums of a finite number of components.

Errors of the type just described have an instructive history, and therefore we shall dwell upon some of its details.

The operations of addition and subtraction in the domain of real numbers are always realizable and single-valued. Mathematicians of the 18th century were well aware of it, at least as applied to rational numbers, but they did not always limit the application of the principles of arithmetic to the domain of sums with a finite number of terms. Along the path of the arbitrary extension of the matter of arithmetical arguments there arose irreconcilable contradictions. The "retention" of single-valuedness for the operation of addition required resorting to different devices, at times clever, yet clearly without adequate basis.

Wishing, for example, to find the "sum" of the series $1 - 1 + 1 - 1 + \ldots$, the calculators wrote $S = 1 - 1 + 1 - 1 + \ldots$, whence, transferring the first term of the right-hand member to the left, they found $S - 1 = -1 + 1 - 1 + 1 - \ldots$, i.e. $S - 1 = -S$, and therefore they concluded that $S = \frac{1}{2}$. But the mathematicians could not overlook other entirely transparent "answers," such as, for example:

$$S = (1 - 1) + (1 - 1) + (1 - 1) + \ldots = 0$$

or

$$S = 1 - (1 - 1) - (1 - 1) - (1 - 1) - \ldots = 1.$$

The question would naturally arise: which solution should be retained as the only true one?

The Italian mathematician Guido Grandi, wishing to establish arguments to confirm the correctness of the answer $S = \frac{1}{2}$, expounded some interesting considerations. Assume, argued Grandi, that some father, on dying, left to his two sons a precious stone. In his will he indicated that after the passage of each year the legacy should pass from one son to the other. Proceeding from these conditions, let us calculate the share of the wealth for each of the heirs. For this, expressing a year of possession of the stone by one of the brothers as $+1$, and the year of possession by the other brother as -1, we obtain the series: $1 - 1 + 1 - 1 + \ldots$, which should be equal to $\frac{1}{2}$, since the brothers have equal rights upon the legacy left to them.

This problem gave rise to a long and lively discussion in which, besides Grandi himself, disputants were Leibnitz, Wolf, Varignon, and Nicolas Bernoulli the elder. Leibnitz showed that the series $1 - 1 + 1 - 1 + \ldots$ bears no relation to the example of Grandi. The brothers will equally share the stone if they keep it, for example, for one hundred years, but the sum of one hundred terms of the series is equal, nevertheless, to zero, and not to a half. However, even Leibnitz considered the answer $\frac{1}{2}$ as the only correct one, supporting its truth by metaphysical arguments about the law of justice in nature.

The equality of the series under consideration to one-half was also maintained by Euler, who asserted that, since with successive summing we obtain now 1 and now 0, and the series has no end, then we should take the mean, i.e. $\frac{0+1}{2} = \frac{1}{2}$.

This curious dispute of the mathematicians of the 18th century was resolved by the establishment of the precise concept of a converging series and the recognition of the fact that a formal extension of the properties of sums with a finite number of components to series is unlawful. The series which served as the object of the discussion is, as is known, a diverging one, i.e. one for which the notion of a sum has no meaning at all.*

37. First of all we note that mathematicians prefer to speak not of the equality of infinite sets, but of their "equivalence": two sets, finite or infinite, are called equivalent, if to each element of

* Here we formulate one of the fundamental assumptions of classical analysis. This has to be noted, since the very question as to the meaning of a given expression is itself meaningful only when the conditions are clearly stated. The point is, that even to the series $1 - 1 + 1 - 1 + 1 - 1 + \ldots$ we may give the single value $\frac{1}{2}$, if we agree to define the sum of a series as $\lim_{n \to \infty} \frac{S_0 + S_1 \ldots + S_n}{n+1}$, where S_n denotes the sum of the first n terms of the series.

the first set there corresponds one, and only one, element of the second; and, conversely, to each element of the second set there corresponds one, and only one, element of the first.

Thus, we have two facts: (1) the set Q is a subset of the set N; (2) the sets N and Q are nevertheless equivalent.

To avoid misunderstanding, we shall note once more that we are dealing with infinite sets, consisting of all the square numbers and all the natural numbers respectively. If we take a finite set of natural numbers not exceeding, for example, one million, and the set of square numbers, also not exceeding one million, then, of course, we shall discover no equivalence whatever between the two sets: to a million natural numbers we shall find about a thousand square numbers, the part will be less than its whole.

Now if we consider all the natural and all the square numbers, without in any way bounding their values, then it will turn out that the set Q, although a part of the set N, at the same time is equivalent to it: the part is equal to its whole.

Having obtained such a conclusion we inevitably raise doubts about the correctness of our reasoning, since we have become accustomed to consider it as one of the basic mathematical facts that a part is less than its whole (this assertion constitutes one of the axioms of Book One of the *Elements* of Euclid). But the reasoning is incontrovertible, and the conclusion has to be accepted. The axiom "the part is less than its whole" is true only for finite sets, while an infinite set may be also equal to its part. To the question, "What sets are called infinite?" we sometimes answer thus: "An infinite set is one in which a part is equivalent to the whole set."

In addition to the algebra of finite numbers there is also an algebra of infinite (or "transfinite") numbers. It is studied in the mathematical discipline which bears the name "Theory of Sets." As we see, both these algebras differ sharply from one another. We have confirmed anew what we said in Problem 35: "If we denote some quantity by the letter x and carry out upon this x operations according to the rules of ordinary algebra, we may be assured of the correctness of the conclusions obtained only if we convince ourselves previously that x is a definite number.

38. Taking instead of a and b any definite numbers, for example, $a = 3$, $b = 1$, it is easy to see that the equalities (1) and (2) are true, while the equalities (3) and (4) are untrue. Consequently, the error is introduced in passing from the equality (2) to the equality (3). This transformation was made on the basis of the consideration that the extraction of a square root of one and the same degree from two equal numbers should lead to equal results. This, of course, is quite correct if we are dealing with single positive numbers: if x and y are two positive numbers, and n is an arbitrary natural number, then from the equality $x^n = y^n$ follows the equality $x = y$. In fact, if $x > y$ or $x < y$, should hold, then also x^n would be greater or less than y^n. For $n = 2$ this theorem may be formulated as follows: if two squares have one and the same area, then their sides are equal.

All this is so if the numbers x and y are positive. But if both (or one) of them may be either positive or negative, then from the equality of the powers of these numbers we may not conclude the equality of the numbers themselves. For example, for $x = 5$ and $y = -5$ we have the equality of the squares, since $x^2 = y^2 = 25$, but here $x > y$.

Taking into account the rule of signs in raising rational numbers to the second power, we easily arrive at the following conclusion: if the squares of two rational numbers are equal, then the numbers themselves are either equal or opposite (i.e. have one and the same absolute value but differing signs). In short: from the equality $x^2 = y^2$ follows one of the equalities $x = y$ or $x = -y$. Which is actually the case has to be clarified every time separately on the basis of the information at our disposal as to the numbers x and y.

In the argument set forth above, which led us to a false conclusion as to the equality of the numbers a and b, we had the equality $(a - b)^2 = (b - a)^2$. If $a > b$, as had been assumed, then $a - b$ is a positive number, and $b - a$ is a negative number.

Thus, here we are dealing with rational numbers and have to take into consideration the fact that from the equality of the squares of these numbers follows one of two things: either these numbers $a - b$ and $b - a$ are equal to each other, or are opposite in sign. But the numbers $a - b$ and $b - a$ have differing signs. The possibility of their equality drops out, and there remains only the possibility of their being opposite in sign. Equality (3) in the argument given above should be replaced by the following:

$$a - b = -(b - a),$$

whence

$$a - b = -b + a, \quad a - a = b - b,$$

and not

$$a + a = b + b,$$

as had been obtained above, and the absurd conclusion is avoided.

To explain the present sophism one often limits oneself simply to indicating the fact that the extraction of a square root gives a result with two signs (plus and minus). For full clarity it is necessary to add a few more considerations. Extracting the square root from both members of equality (2), we should write the result in the form $\pm(a - b) = \pm(b - a)$, where any sign of the left-hand member may be taken simultaneously with any sign of the right-hand member. Instead of the one equality (3) we now have four equalities:

$$+(a - b) = +(b - a),$$
$$+(a - b) = -(b - a),$$
$$-(a - b) = +(b - a),$$
$$-(a - b) = -(b - a).$$

Changing the signs of both members in the last two equalities for the opposite signs, we reduce these four equalities to two:

$$+(a - b) = +(b - a),$$
$$+(a - b) = -(b - a),$$

and making use of the double sign $\pm$, even to the single:

$$a - b = \pm(b - a),$$

in which one still has to carry out the choice of sign.

Mistakes caused by passing from $x^2 = y^2$ to $x = y$ are met very frequently. It is necessary to remember strictly that from $x^2 = y^2$ there follow two equalities: $x = y$ and $x = -y$, or, in short, $x = \pm y$, and never to forget the necessity to clarify the choice of sign.

39. It is equality (2) which is erroneous: from equality (1) it follows that either $x - 2a = x$, or $x - 2a = -x$. On the strength of the condition $x = a$ we have: $x - 2a = a - 2a$, and therefore equality (2) drops out (for $a \neq 0$ the number $x - 2a$ equal to $-a$ cannot be equal to x, which is equal to a) and should be replaced by the other: $x - 2a = -x$. But from this it follows that $2x - 2a = 0$ and $x = a$. We have arrived at the original equality, having eliminated the absurd conclusion.

40. Just as in Problem 39 the mistake is introduced in the extraction of the square root from both members of equality (1): from equality (1) it follows that the expressions in brackets are either equal, and then equality (2) holds, or these expressions are equal in absolute value but opposite in sign, and then instead of (2) we should write the equality:

$$n - \frac{2n+1}{2} = -\left(n + 1 - \frac{2n+1}{2}\right). \tag{3}$$

It is easy to convince ourselves of the correctness of equality (3) and not that of (2).

41. Let us see how the points A, B, B_1, B_2, B_3, B_4, and so on, with which we are dealing, are situated, and how much time is necessary for the point M to traverse the segments AB, BB_1, B_1B_2, B_2B_3, and so on. According to the condition, $AB = 100$ m. The segment BB_1 is traversed by the point N at a speed of 1 m/sec during a period in the course of which the point M, moving at a speed of 10 m/sec, traverses a segment $AB = 100$ m, i.e. in

$100 : 10 = 10$ (sec). It is easily seen that $BB_1 = 10$ m. In exactly the same way we establish that in order to traverse from B to B_1, the point M requires 1 sec (during which the point N will traverse the segment $BB_2 = 1$ m), and to traverse from B_1 to B_2, 0·1 sec is necessary (in that period the point N will traverse the segment $B_2B_3 = 0{\cdot}1$ m). Repeating these arguments, it is easy to convince oneself that the distances (in metres) between the points A and B, B and B_1, B_1 and B_2, and so on form an infinitely decreasing geometric progression:

$$100;\ 10;\ 1;\ 0{\cdot}1;\ 0{\cdot}01;\ 0{\cdot}001;\ \ldots,$$

and the segments of time (in seconds) during which the point M traverses these distances form another such progression:

$$10;\ 1;\ 0{\cdot}1;\ 0{\cdot}01;\ 0{\cdot}001;\ \ldots$$

The argument given above, "proving" that the point M will never reach the point N, in effect proves that the point M will not reach the point N in the course of a period of time equal to the sum of any number n of the segments 10, 1, 0·1, 0·01, 0·001, . . . But the sum of n terms of that progression is equal to:

$$S_n = \frac{10 - 10 \times 0{\cdot}1^n}{1 - 0{\cdot}1} = \frac{10}{1 - 0{\cdot}1} - \frac{10 \times 0{\cdot}1^n}{1 - 0{\cdot}1}$$

and is always less than the number $\frac{10}{1 - 0{\cdot}1} = \frac{10}{0{\cdot}9} = 11\frac{1}{9}$. Consequently, our argument proves only that the point M will not reach the point N in any period of time less than $11\frac{1}{9}$ sec. In that assertion there is nothing absurd: the point M approaches the point N by $10 - 1 = 9$ (m/sec) and will reach it, i.e. will become closer by the initial distance between them of 100 m, in precisely $100 : 9 = 11\frac{1}{9}$ sec, but not earlier.

The sophism just analysed was pointed out by the Greek philosopher Zeno.

In ancient Greek mathematics, among the basic axioms there was the assertion that "the sum of an infinitely great number of any, however small, extended quantities is necessarily infinitely great."* This feature is pointed out by the commentator of Aristotle, Simplicius (died in 549).

With the admission of this axiom are connected the famous "paradoxes"—the proofs of Zeno the Elean, who lived in the fifth century B.C. Diogenes Laertius (a writer of the end of the second and beginning of the third centuries) testifies that Zeno possessed an unusual gift of rhetoric, wrote works showing the great power of his mind and profound learning, and became known in philosophy and politics. In philosophy his name is connected with the negation of the possibility of expressing motion in a scientific formulation, and in politics with an active support of the forces of reaction in an open political struggle against the ancient democracy.

Zeno spoke on behalf of the opposition of a perfectly definite mathematical theory according to which one simultaneously postulated the existence of a least, further indivisible, segment of space or time, and the infinite divisibility of quantities. To his sceptical views the ancient philosopher added the brilliant form of splendidly constructed indirect proof.

"There are four of Zeno's arguments on motion ('Dichotomy', 'Achilles and the Tortoise', 'The Arrow', and 'Ristallsis')," says Aristotle, "which pose considerable difficulties to those who wish to solve them." "The second," continues Aristotle, "is the so-called Achilles. It consists in that an object slower in motion will never be overtaken by the very fastest, since the pursuer has first to reach the place whence the escapee has already moved, so that the slower one always has a certain advantage."†

The psychological foundation of the above paradox is the intuitive identification of the sum of an infinite number of terms with an infinitely great quantity, even though in fact the matter

* S. Ya. Lur'ye, *Theory of the Infinitesimal as per the Ancient Atomists* (Teoriya beskonechno malykh u drevnykh atomistov), p. 31, Moscow, 1935.

† Aristotle, *Physics* (Fizika), transl. by V. P. Karpov, pp. 119–120, Moscow, 1936.

deals only with the sum of an infinitely decreasing geometric progression $1 + \frac{1}{n} + \left(\frac{1}{n}\right)^2 + \left(\frac{1}{n}\right)^3 + \ldots$, equal to the fraction $\frac{n}{n-1}$, where n indicates how many times faster Achilles moves than the tortoise.

The level of philosophic and mathematical culture in the Greece of Zeno's time does not allow us to assume that such gross mistakes could lead the outstanding thinkers of that epoch into error. To note that the sum of an arbitrarily great number of terms of such a series will, nevertheless, not exceed some number could not constitute a great difficulty for them. The whole point is that the Greek scholar approached the solution of this problem, proceeding from the unqualified acceptance of two contradictory points of view: the unlimited divisibility and the existence of the least elements of division. The actual sense of the "Achilles" consists precisely in illustrating the fact that the sum of an infinitely great number of indivisibles is always an infinitely great quantity.

Upon such an interpretation of Zeno's paradox the authoritative historians of mathematics converge. This hypothesis was first put forth by P. Tannery. The famous Soviet student of Hellenism, S. Ya. Lur'ye, sharing the view of Zeller, Nestle, and Hiss, disputes Tannery's opinion that in Zeno's opponents one should see the Pythagoreans. Lur'ye indicates that "all that we know of Anaxagoras speaks for the fact that it was he who from the ancient point of view was accused of precisely that contradiction on which Zeno was harping."* However, Professor Lur'ye does not assert that Zeno had in view precisely Anaxagoras (500–428), considering that the contemporary state of science still does not give sufficient basis for solving this question.

This, then, is the "Achilles" of Zeno against the background of the mathematics and philosophy contemporary to him.

But Aristotle deprived Zeno's paradoxes on motion of the basis

* S. Ya. Lur'ye, *loc. cit.*, p. 34.

from which they arose. He accepted the infinite divisibility of continuous quantities and denied the existence of least elements of division (indivisibles).* Because of this, according to Aristotle, the sum dealt with in the "Achilles" is no longer an infinitely great quantity.

The Aristotelian explanation of Zeno's paradox has deprived it of mathematical basis. However, the discussion of the original gnosiological sources of Zeno's paradox did not leave the pages of historico-philosophical literature for the course of many centuries. Only dialectics allowed to demonstrate finally the flimsiness of Zeno's philosophy and in particular, of his paradoxes. V. I. Lenin, in laying bare the essence of Zeno's paradoxes, emphasized that space is divided in this case abstractly, in the brain, while, in fact, i.e. in the process of motion, it is both divisible and indivisible at the same time, since real motion represents a contradictory unity of continuity and discontinuity of space and time.†

42. I. $$\frac{a+b}{ac} = \frac{1+b}{c}. \qquad (3)$$

A very widespread error: the simplification by a common factor of the denominator with one of the components of the numerator.

Examination yields: $ac + bc = ac + abc$; $bc = abc$ or $b = ab$, since the denominator c cannot be equal to zero. From the relation $b(1 - a) = 0$ it follows that the equality (3) will be correct only in two special cases: firstly, if $b = 0$ and the formula (3) takes the form $\frac{a}{ac} = \frac{1}{c}$; second, when $a = 1$ and the formula (3) takes the form $\frac{1+b}{c} = \frac{1+b}{c}$.

* Aristotle, *loc. cit.*, p. 105.

† V. I. Lenin, *Philosophical Notebooks* (Filosofskie tetradi), pp. 242–243, Moscow, 1947.

$$\text{II.}\quad \frac{a}{b+c} = \frac{a}{b} + \frac{a}{c}.$$

A typical and very stubborn mistake.

Examination yields:

$$abc = ac(b+c) + ab(b+c);\quad abc = abc + ac^2 + ab^2 + abc;$$
$$ac^2 + ab^2 + abc = 0;\quad a(b^2 + c^2 + bc) = 0.$$

From the last equality it follows that either $a = 0$, or

$$b^2 + c^2 + bc = 0.$$

However, the trinomial $b^2 + c^2 + bc$ will not vanish for any real values of b and c other than zero. In fact, if $b \neq c$, then either $b^2 > |bc|$ or $c^2 > |bc|$; if $b = c$, then $b^2 = c^2 = bc$. And therefore in all the cases

$$b^2 + c^2 + bc > 0.$$

It comes out that "formula" (4) proves to be correct only in the trivial case when $a = 0$.

It is not by accident that we have dwelt on examples referring to identical transformations over algebraic fractions and to operations upon them. The point is that a very great number of mistakes are made by students in this field.

$$\text{III.}\quad \sqrt{a^2 + b} = a\sqrt{b}. \tag{5}$$

Mistakes of this type are very widespread and require stubborn efforts if they are to be prevented.

Investigation yields:

$$a^2 + b = a^2b;\quad a^2 = b(a^2 - 1);\quad b = \frac{a^2}{a^2 - 1}. \tag{6}$$

Formula (5) holds in the set of real numbers, when $a > 1$ and the relation (6) is fulfilled.

If, for example, $a = 2$, then $b = \frac{4}{3}$ and

$$\sqrt{(4 + \tfrac{4}{3})} = 2\sqrt{(\tfrac{4}{3})}.$$

CHAPTER IV

Geometry

I. Examples of False Arguments

43. Segments of parallel straight lines bounded by the sides of an angle are equal

Take an arbitrary angle and intersect its sides by two arbitrary parallel straight lines. Let AB and CD be the segments of the parallels included between the sides of that angle, and E its vertex (Fig. 16).

As is well known, parallel straight lines intersect proportional segments on the sides of the angle. Consequently,

$$AE : CE = BE : DE$$

and

$$AE \times DE = BE \times CE. \tag{1}$$

Multiplying both members of equality (1) by the difference $AB - CD$, we carry out the following transformations:

$$AE \times DE \times AB - AE \times DE \times CD$$
$$= BE \times CE \times AB - BE \times CE \times CD,$$

$$AE \times DE \times AB - BE \times CE \times AB$$
$$= AE \times DE \times CD - BE \times CE \times CD,$$

$$AB(AE \times DE - BE \times CE)$$
$$= CD(AE \times DE - BE \times CE) \tag{2}$$

Dividing both members of the last equality by the difference $AE \times DE - BE \times CE$, we obtain the equality $AB = CD$. Thus the segments of parallels confined between the sides of a given angle are always equal.

44. Segment of a straight line is equal to a proper part of itself

Intersect an arbitrary straight line at the points A and B by the straight lines MN and PQ, perpendicular to AB (Fig. 17).

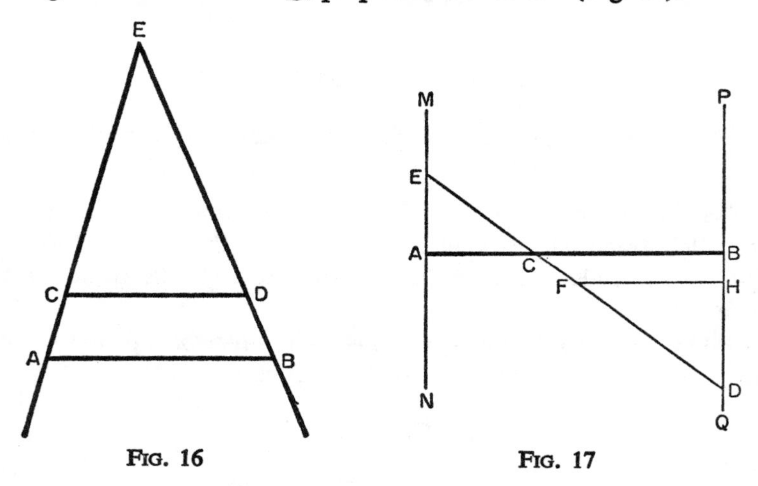

FIG. 16 FIG. 17

Draw a straight line intersecting MN, AB, and PQ at the points E, C, and D, respectively.

From the similarity of the triangles CBD and ACE we establish:

$$\frac{BD}{AE} = \frac{CB}{AC} \quad \text{or} \quad \frac{BD}{AE} = \frac{AB - AC}{AC}. \tag{1}$$

At the points F and H, intersect the sides CD and BD of the triangle CBD by a straight line parallel to its third side.

From the similarity of the triangles CBD and FHD we establish:

$$\frac{BD}{HD} = \frac{BC}{FH}, \quad \text{or} \quad \frac{BD}{HD} = \frac{AB - AC}{FH}. \tag{2}$$

Determining BD from the relations (1) and (2) we have, respectively:

$$BD = \frac{AE \times (AB - AC)}{AC};$$

$$BD = \frac{HD \times (AB - AC)}{FH}.$$

Consequently,

$$\frac{AE \times (AB - AC)}{AC} = \frac{HD \times (AB - AC)}{FH}. \qquad (3)$$

Eliminating the denominators and removing the brackets, we obtain:

$$AE \times FH \times AB - AE \times FH \times AC$$
$$= AC \times HD \times AB - AC^2 \times HD.$$

Adding to both members of the last equality the difference $AE \times HF \times AC - DH \times AB \times AC$, after collecting terms and removing the common factors, we have:

$$AB \times (AE \times FH - AC \times HD)$$
$$= AC \times (AE \times FH - AC \times HD) \qquad (4)$$

Finally, from relation (4) we establish:

$$AB = AC. \qquad (5)$$

45. All triangles are of equal area

Designate the sides of an arbitrary triangle by the letters a, b, and c, the corresponding altitudes—h_1, h_2, and h_3, the area—S.

In using other arbitrary triangles we shall keep to the same notation but in order to distinguish between them we shall use primed letters.

From the geometry course it is known that:

$$\frac{S}{S'} = \frac{ah_1}{a'h'_1}; \tag{1}$$

$$\frac{S}{S''} = \frac{bh_2}{a''h''_1}. \tag{2}$$

Determining S from the relations (1) and (2) respectively, we have:

$$S = S' \times \frac{ah_1}{a'h'_1}; \qquad S = S'' \times \frac{bh_2}{a''h''_1}.$$

Consequently,

$$S' \times \frac{ah_1}{a'h'_1} = S'' \times \frac{bh_2}{a''h''_1}$$

or, eliminating the denominators:

$$S'ah_1a''h''_1 = S''a'h'_1bh_2. \tag{3}$$

Multiplying both members of equality (3) by the difference $S - S'$ and removing brackets, we have:

$$SS'ah_1a''h''_1 - S'^2ah_1a''h''_1 = SS''a'h'_1bh_2 - S'S''a'h'_1bh_2. \tag{4}$$

Adding to both members of equality (4) the difference $S'^2ah_1a''h''_1 - SS''a'h'_1bh_2$, after collecting like terms and removing the common factors, we obtain:

$$S(S'ah_1a''h''_1 - S''a'h'_1bh_2) = S'(S'ah_1a''h''_1 - S''a'h'_1bh_2). \tag{5}$$

Finally, from the relation (5) we establish:

$$S = S'. \tag{6}$$

46. The sum of the bases of an arbitrary trapezoid is equal to zero

To prove this astonishing "theorem" we need a property of a set of equivalent ratios, known from the elementary algebra course: if $\frac{p}{q} = \frac{r}{s}$ then each ratio is equal to $\frac{p - r}{q - s}$.

Take an arbitrary trapezoid $ABCD$ (Fig. 18) and extend the lower base a, say to the right, by a segment equal to the upper base b. The upper base b we shall extend in the opposite direction, i.e. to the left, by a segment equal to the lower base a. Draw the diagonals of the trapezoid and designate by the letters x, y, z the three segments, into which the diagonal AC is cut by the other diagonal BD and by the straight line joining the ends of the extensions of the bases, i.e. the straight line FE.

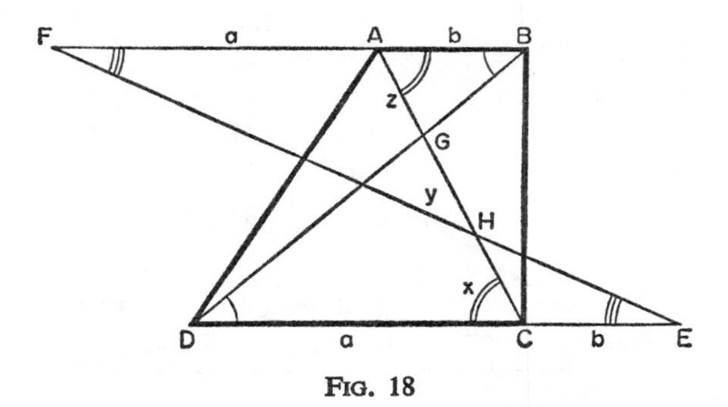

FIG. 18

From the similarity of the triangles CDG and ABG, we have the proportion $(x + y) : z = a : b$, and from the similarity of the triangles AFH and CEH we have the proportion $(y + z) : x = a : b$, which yields a new proportion $(x + y) : z = (y + z) : x$. Multiplying both members of the second proportion by -1, we arrive at the proportion

$$(x + y) : z = (-y - z) : (-x).$$

Applying to that proportion the above-mentioned property of a set of equal ratios, we shall obtain, that

$$(x + y) : z = (x + y - y - z) : (z - x)$$
$$= (x - z) : (z - x) = -1,$$

i.e.

$$(x + y) : z = -1.$$

Combining the result obtained with the proportion

$$(x + y) : z = a : b,$$

we find that $a : b = -1$, whence $a = -b$ and $a + b = 0$.

47. Inscribed and circumscribed

In Fig. 19 we have two broken lines with common ends, the inscribed line ADC and the circumscribed line ABC. As is well known, a convex inscribed line is always shorter than its circumscribed line. The following argument leads to a result which is in contradiction with that theorem.

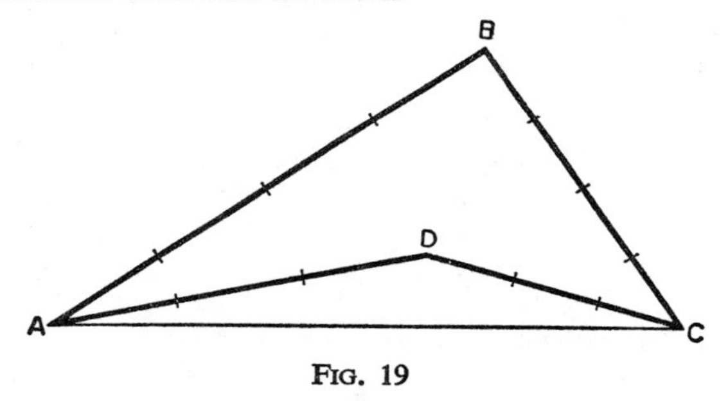

FIG. 19

Let the segments AB and BC be taken arbitrarily, and the segments AD and DC be taken not arbitrarily but as proportional to the segments AB and BC, i.e. $AD = k \times AB$, $DC = k \times BC$, where k is some proper positive fraction (in Fig. 19 we have taken $k = \frac{3}{4}$). From the given equalities it follows that

$$-AD = k(-AB),$$

$$-DC = k(-BC),$$

$$(-AD) + (-DC) = k[(-AB) + (-BC)],$$

$$[(-AD) + (-DC)] : [(-AB) + (-BC)] = k,$$

but

$$k = AD : AB,$$

consequently,

$$[(-AD) + (-DC)] : [(-AB) + (-BC)] = AD : AB. \quad (1)$$

In the proportion (1) the first term of the second ratio is less than its second term. Consequently the first term of the first ratio is also less than its second term, i.e.

$$(-AD) + (-DC) < (-AB) + (-BC). \qquad (2)$$

Transferring all the terms from the first member of the inequality to the second and from the second to the first, we have:

$$AB + BC < AD + DC.$$

Thus, the circumscribed line ABC turns out to be not longer, as should be, but shorter than its convex inscribed line ADC.

48. More about proportionality

In Problem 7 (Chap. II) we had a few examples of mistakes introduced by taking as proportional quantities which are in no way proportional. Here are examples of errors in problems of a geometric character, caused by the same effect.

I. On a plan with a scale of 1 : 10,000 a rectangle is represented which on the plan has sides of 2 cm and 3 cm. What is the area of that rectangle in nature? Here the fraction 1 : 10,000 ("numerical scale") is the ratio of the length of any segment on the plan to the length of the corresponding segment in nature.

Is the assertion that the area of the rectangle in nature is $(2 \text{ cm} \times 3 \text{ cm}) \times 10{,}000 = 60{,}000 \text{ cm}^2 = 6 \text{ m}^2$ correct?

II. There are three drums in the shape of circular cylinders. The second has a base diameter twice as large as the first, but on the other hand its altitude is one-half as large. The third drum has a base diameter one-half the length of that of the first drum, while its altitude is twice as large. Is it correct to say that the doubling of one dimension compensates for the halving of the other and that therefore the volume of all the three drums is one and the same?

III. The model of a structure in $\frac{1}{20}$ natural size is made of the same material as the structure itself. Weighing has shown that this model weighs 3 kg. Should one expect that the structure itself will weigh 3 kg $\times$ 20 = 60 kg?

IV. Wishing to compare two plots, a man has measured their boundaries, and finding the boundary of the first plot equal to 60 m and the boundary of the second equal to 50 m, concluded that the area of the first plot is greater than the area of the second one by 20 per cent. Is it really so?

49. The circumferences of two circles of different radii have one and the same length

Take two wheels with radii R and $r < R$ and imagine that they are set on a common axis. Roll the wheel of radius R without sliding along the straight line DE (Fig. 20). When the point A on the circumference of that wheel, initially on the straight line DE, has undergone a complete revolution and is again on the straight line DE, coinciding with the point A_1, then the path CC_1, traversed in that time by the centre of the circle, is equal to the segment AA_1, which is equal, in its turn, to the length of the circumference of the wheel $2\pi R$.

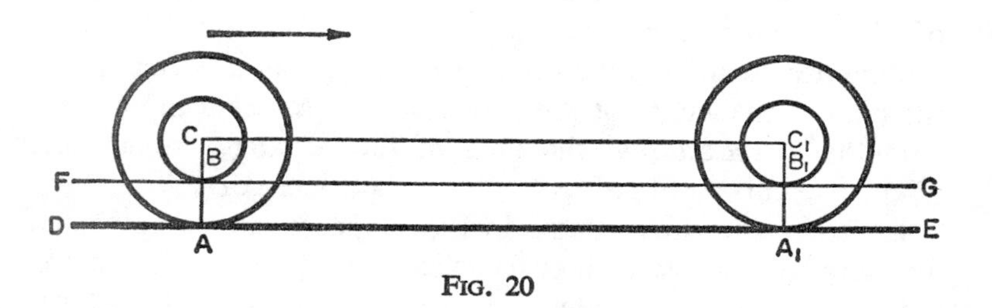

FIG. 20

If the second (smaller) wheel is set on a common axis with the first and rigidly fixed to it, then both wheels undergo one complete revolution at the same time. But one may consider that while the first wheel rolls along the straight line DE, the second wheel rolls along the straight line FG. In undergoing one complete revolution the second wheel will traverse the path BB_1, equal to the length of its circumference, i.e. $2\pi r$. But $BB_1 = CC_1$, and therefore $2\pi r$, i.e. the lengths of the two circumferences of differing radii turn out to be equal!

Where is the error in our argument?

50. The sum of the arms is equal to the hypotenuse

Take an arbitrary right-angled triangle ABC and divide its hypotenuse into n equal parts, where n is some natural number, and then through every point of division draw a pair of rectilinear segments, one parallel to the arm AB, the other parallel to the arm AC. Extending these segments up to their mutual intersections outside the triangle, we obtain a steplike broken line,

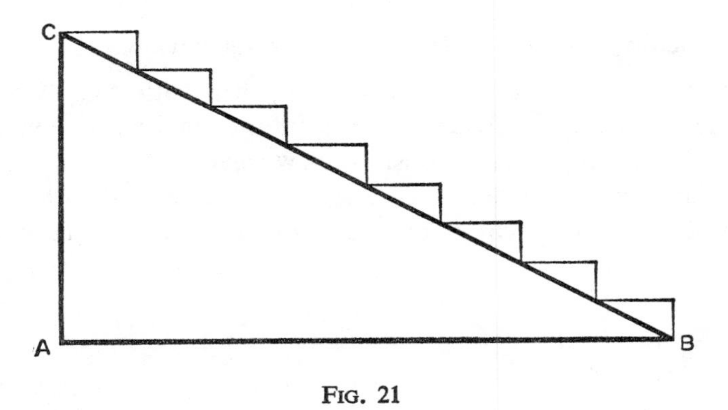

FIG. 21

represented in Fig. 21. The sum S_n of all the links of that broken line from point B to point C is equal to the sum of the arms $AB + BC$, since the sum of all the segments drawn parallel to one of the arms is equal to that arm.

We now increase without bound the number n, giving it successively the values 2, 4, 8, 16, . . . and so on. The number of links in our broken line BC will then increase without limit (it is equal to $2n$), but the length of each link will tend to zero and the broken line will differ ever less and less from the straight line BC.

In the limit, as $n \to \infty$, the broken line will coincide with the hypotenuse BC, and therefore

$$\text{limit } S_n = BC. \tag{1}$$

But for any natural number n we always have, as we had seen above, the equality $S_n = AB + AC$. Consequently, also the preceding S_n as $n \to \infty$ is equal to the same sum:

$$\text{limit } S_n = AB + AC. \tag{2}$$

The combination of the equalities (1) and (2) leads to the conclusion that $AB + AC = BC$, i.e. that the sum of the arms of an arbitrary right-angled triangle is equal to its hypotenuse.

51. The length of a semicircle is equal to its diameter

Take a semicircle of radius r together with the diameter bounding it and we divide the latter into n equal parts. On each segment of the diameter, construct new small semicircles, placing them in turn on one and then on the other side of the diameter. We obtain a wavy curve, curling about the diameter (Fig. 22).

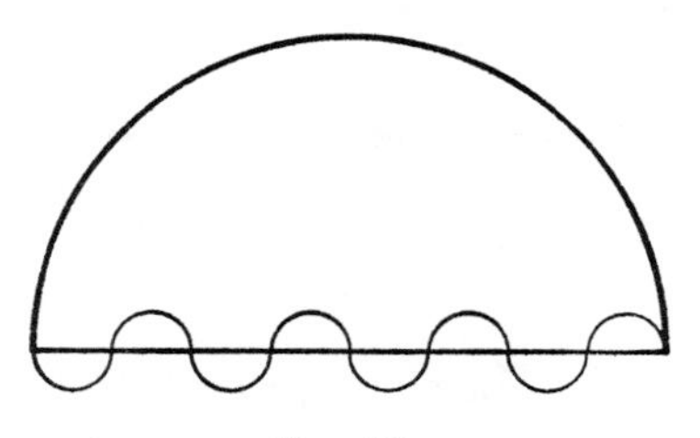

FIG. 22

As the number n increases without limit, this curved line becomes ever less and less different from a straight line and in the limit, as $n \to \infty$, coincides with it. Denoting the length of that curved line composed of n equal semicircles by the letter L_n, we therefore have:

$$\text{limit } L_n = 2r. \tag{1}$$

But the length of every small segment of the semicircle, constructed on diameter $\frac{2r}{n}$ and with radius $\frac{r}{n}$, is equal to $\left(2\pi \times \frac{r}{n}\right) : 2 = \frac{\pi r}{n}$, and therefore $L_n = \frac{\pi r}{n} \times n = \pi r$. Thus, the complete length of that curved line:

$$\text{limit } L_n = \pi r. \tag{2}$$

Combination of the equalities (1) and (2) leads to the conclusion that $\pi r = 2r$, $\pi = 2$. Consequently, the length of an arbitrary semicircle is equal to the length of its diameter and the number π, expressing the ratio of the length of any circumference to its diameter, is equal to 2.

52. The lateral surface of a right circular cone with base radius r and altitude h is expressed by the formula $P = \pi r(r + h)$

Take a right circular cone with a base radius r and altitude h, and in it make sections by planes perpendicular to the axis of the cone, distant from each other by $\frac{h}{n}$, where n is some natural number. Each of the $n - 1$ circular sections obtained we take as the upper base of a cylinder having as altitude the segment $\frac{h}{n}$. In all we obtain $n - 1$ cylinders, forming together a *steplike* figure inscribed in the given cone. The cross-section of the cone and of that steplike figure along the axis of the cone is represented in Fig. 23.

Let us find the side surface of the body. It consists of the sum of the curved surfaces of all the cylinders and of the sum of the surfaces of the rings remaining on the upper base of each cylinder, after subtracting the area occupied by the lower base of the next upper cylinder. On the very top cylinder the upper base is to be taken whole.

Having established from similarity of triangles that the radius of the base of the lowest cylinder is equal to $r\left(1 - \frac{1}{n}\right)$, of the second from bottom $r\left(1 - \frac{2}{n}\right)$, third from bottom $r\left(1 - \frac{3}{n}\right)$ and so forth, up to the very topmost, in which the radius of the

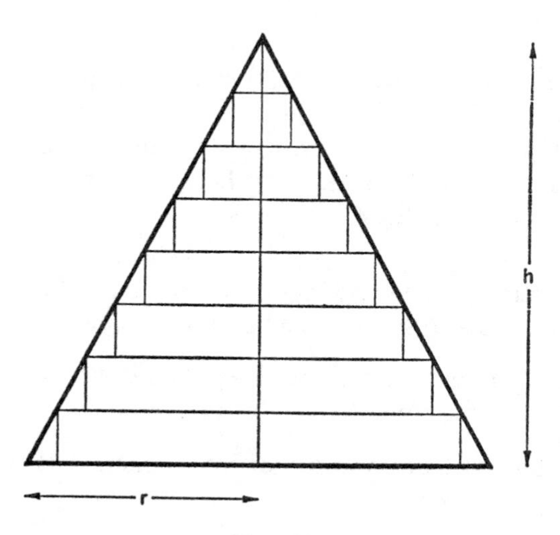

FIG. 23

base is equal to $r\left(1 - \frac{n-1}{n}\right)$, we easily find that the sum of the side surfaces of all the cylinders is equal to:

$$2\pi r\left(1 - \frac{1}{n}\right) \times \frac{h}{n} + 2\pi r\left(1 - \frac{2}{n}\right) \times \frac{h}{n} + 2\pi r\left(1 - \frac{3}{n}\right) \times \frac{h}{n}$$

$$+ \ldots + 2\pi r\left(1 - \frac{n-1}{n}\right) \times \frac{h}{n},$$

or $$2\pi r \times \frac{h}{n}\left[\left(1-\frac{1}{n}\right)+\left(1-\frac{2}{n}\right)+\left(1-\frac{3}{n}\right)+\dots + \left(1-\frac{n-1}{n}\right)\right]$$

or

$$2\pi r \times \frac{h}{n}\left[(n-1)-\frac{1}{n}\times\frac{n(n-1)}{2}\right]=\pi rh\left(1-\frac{1}{n}\right).$$

As to the area of the rings, their sum is equal to the area of the base of the lowest (the largest) cylinder, i.e. $\pi\left[r\left(1-\frac{1}{n}\right)\right]^2$. For the side surface of the steplike body we obtain the formula:

$$P_n=\pi rh\left(1-\frac{1}{n}\right)+\pi\left[r\left(1-\frac{1}{n}\right)\right]^2,$$

or, upon simplification:

$$P_n=\pi r\left[h\left(1-\frac{1}{n}\right)+r\left(1-\frac{1}{n}\right)^2\right].$$

Noting that as one increases the number n without bound the steplike body will approach a cone, at limiting value we find for the curved surface P of the cone the formula indicated in the title:

$$P=\text{limit}\,P_n=\text{limit}\,\pi r\left[h\left(1-\frac{1}{n}\right)+r\left(1-\frac{1}{n}\right)^2\right],$$

$$P=\pi r(h+r).$$

But this formula differs considerably from that derived in any course of geometry ($P=\pi rl$, where l is the slant height of the cone), since the sum of the segments h and r are the legs of the right triangle, always greater than its hypotenuse l.

53. At a given point on a straight line two perpendiculars can be constructed to that line

On the straight line MN the point A is given. Construct at the point A the perpendicular AK to that straight line.

Draw now a circle of arbitrary radius, passing through A. Denoting by the letter B the second point of intersection of the circumference with the straight line MN we draw the diameter BC. Finally we shall join the points A and C.

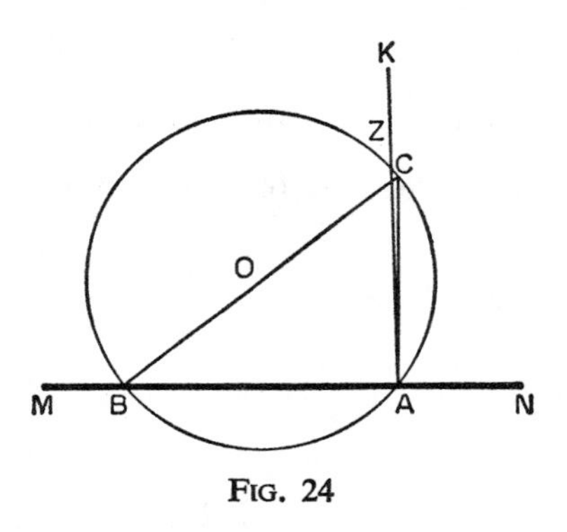

FIG. 24

AC is perpendicular to MN: the angle BAC, as an inscribed angle standing upon the diameter, is equal to 90°.

Thus, at the point A, two perpendiculars to the straight line MN are constructed: AK and AC.

54. Through one point it is possible to draw two straight lines parallel to a given straight line

Say MN is the given line, and K is an arbitrary point not lying on that line.

Through the point K draw the straight line PQ parallel to the straight line MN. Joining the point K to the arbitrary point L on the straight line MN, describe a semicircle on the segment KL as the diameter. Finally, at the point L we construct a perpendicular intersecting the arc of the semicircle at some point E.

The straight line *KE* is parallel to *MN*. In fact, the angle *KEL*, as an inscribed angle subtended by the diameter, is equal to 90°.

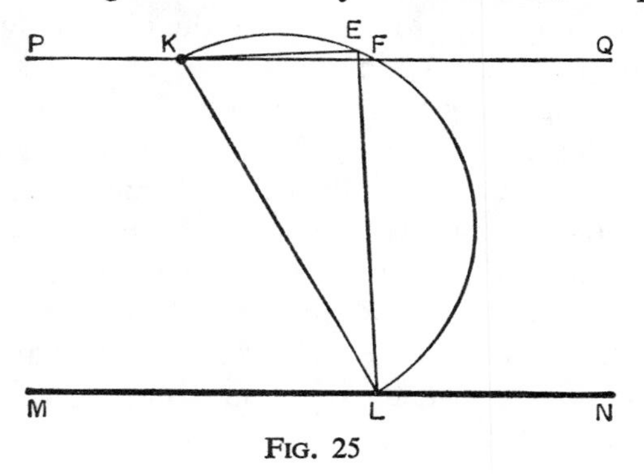

FIG. 25

Thus, to the straight line *MN* two parallel lines are drawn through the point *K*: *PQ* and *KE*.

55. A circle has two centres

Construct an arbitrary angle *ABC* and, taking on its sides two arbitrary points *D* and *E*, construct through them perpendiculars to the sides of the angle (Figs. 26*a* and 26*b*). These perpendiculars

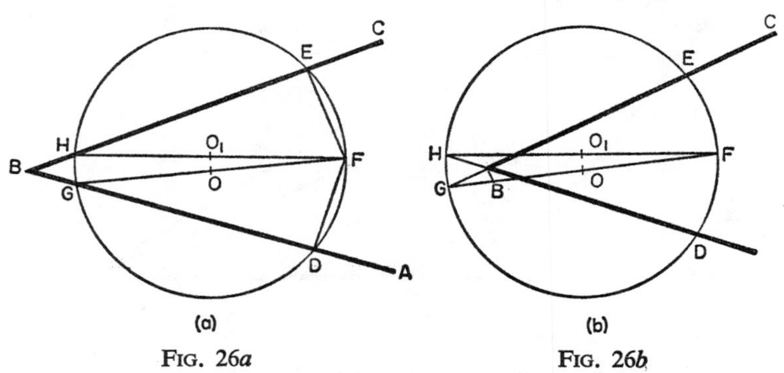

FIG. 26*a* FIG. 26*b*

must intersect (if they were parallel, then also the sides *AB* and *CB* would have been parallel). We denote this point of intersection by the letter *F*.

Through the three points D, E, F draw a circle; this is always possible, since these three points do not line on one straight line. If the point B turns out to be outside the circle (Fig. 26*a*), then after joining the points of intersection H and G to the point F we obtain two right angles GDF and HEF inscribed in the circle. Hence we conclude, that each of the arcs $GHEF$ and $HGDF$ is equal to a semicircle (an inscribed angle is measured by one-half the arc on which it stands), and therefore the segments GF and HF are diameters of our circle. Consequently, the points O and O_1, dividing the segments GF and HF in half, represent precisely the two centres of that circle. The assumption that the point B is inside the circle drawn through the points D, E, F (Fig. 26*b*) leads to the same conclusion.

56. Two perpendiculars may be dropped from a point to a straight line

Take a triangle ABC and on its sides AB and BC as diameters, draw the circles I and II (Fig. 27). Join the points D and E of intersection of these circles with the side AC to the point B. The

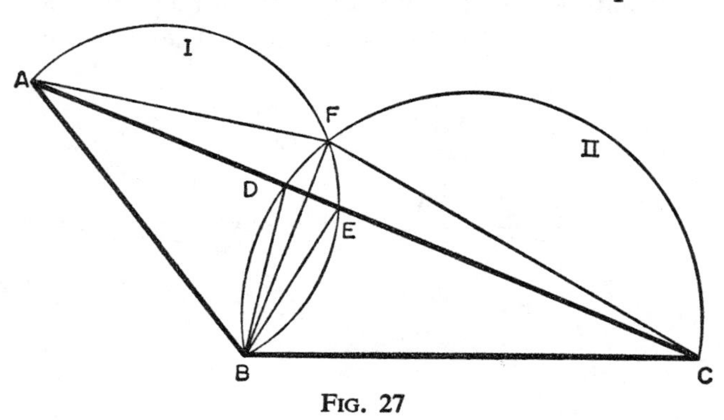

FIG. 27

angle BEA, being an inscribed angle in the circle I and subtended by the diameter, is a right angle, and therefore $BE \perp AC$. The angle BDC, inscribed in the circle II and subtended by its diameter, is also a right angle, and consequently, $BD \perp AC$. Thus, from the point *B* two perpendiculars are drawn to the line *AC*—*BE* and *BD*.

57. Two straight lines may be drawn through two given points

Take an arbitrary triangle *ABC* (Fig. 28) and on each of its sides *AB* and *AC*, as diameters, construct a circle. The two circles, intersecting at the point *A*, will intersect also at some

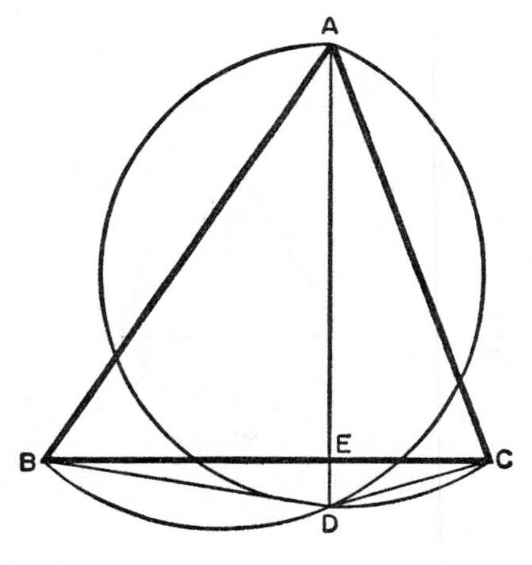

FIG. 28

other point *D*. The angle *ADB*, as inscribed in the circle and subtended by its diameter *AB*, is a right angle. For the same reason the angle *ADC* subtended by the diameter *AC* is also a right angle. The angles *ADB* and *ADC*, having a common vertex *D*, a common side *AD*, and constituting together two right angles, have two other sides *BD* and *DC* on one straight line. Consequently, the line *BDC* is not broken, as shown on the diagram, but is straight. Thus, on the diagram there are two straight lines connecting the points *B* and *C*: the straight line *BDC* and the straight line *BEC*.

58. Every triangle is isosceles

Take an arbitrary triangle ABC and assume that $AC > BC$. Draw the bisector CC_1 of the angle ACB and the axis of symmetry DD_1 of the side AB (the axis of symmetry of a segment we call the straight line dividing the segment into halves and perpendicular to it). These two lines CC_1 and DD_1 can neither coincide with each other, nor be parallel, since in both these cases the bisector would serve at the same time as the altitude, which is

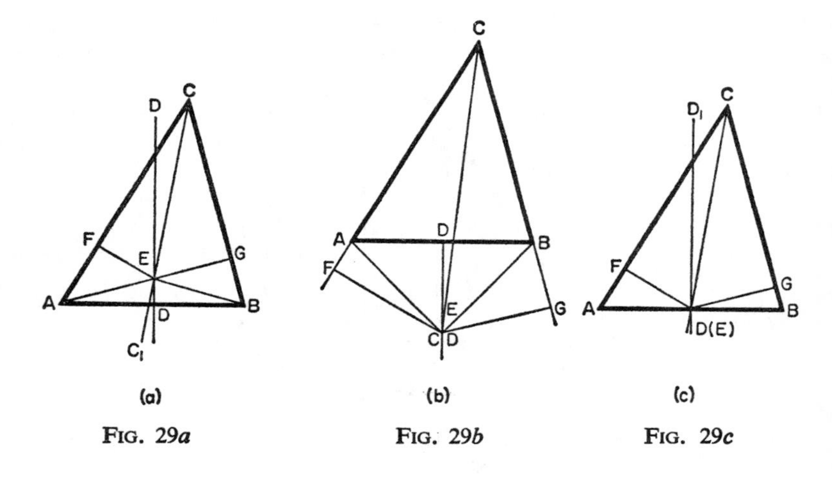

FIG. 29*a* FIG. 29*b* FIG. 29*c*

possible only in an isosceles triangle, i.e. if $AC = BC$. Consequently, the straight lines CC_1 and DD_1 necessarily intersect at some point E. With respect to that point E, three assumptions are possible: either the point E is within the triangle ABC, or outside it, or on the side AB.

Figure 29*a* corresponds to the first assumption. Draw segments AE and BE, and also $EF \perp AC$ and $EG \perp BC$. The right-angled triangles CEF and CEG have equal hypotenuses CE and arms $EF = EG$ (the point E, being on the bisector of the angle C is equidistant from its sides), and therefore $CF = CG$. The right-angled triangles ADE and BDE are also congruent, by virtue of the common arm DE and equal arms $AD = BD$, and therefore $AE = BE$. Finally, the right-angled triangles AEF and BEG are congruent, by virtue of the equality of the hypotenuses AE and BE and of the arms EF and EG. Consequently, $AF = BG$. Term-by-term addition of the equalities $CF = CG$ and $AF = BG$ leads to the equality $AC = BC$, contradicting the condition $AC > BC$. We have found that every non-isosceles triangle is at the same time isosceles!

We arrive at the same conclusion, if we make the assumption that the point E is not within, but outside the triangle ABC (Fig. 29*b*). The analysis of the triangles EFC and EGC, EAD and EBD, EAF and EBG allows us to establish that $CF = CG$, $AF = BG$, and the term-by-term subtraction of the last equalities leads to the conclusion that $AC = BC$.

Nothing is changed by the assumption that the point E is on the side AB, i.e. coincides with the point D (Fig. 29*c*). The congruence of the triangles ADF and BDG, CDF and CDG implies the equalities $AF = BG$, $FC = GC$, whence again $AC = BC$.

Thus, with each of the three assumptions made we reach one and the same absurd conclusion. Wherein does the lapse lie?

59. The arm of a right-angled triangle is equal to its hypotenuse

In the right-angled triangle ABC (Fig. 30) BO is the bisector of the angle B, D is the mid-point of the arm AC, $DO \perp AC$, $OE \perp AB$, $OF \perp BC$.

It is easy to see that the right-angled triangles BOE and BOF are congruent, as they have a common hypotenuse BO and equal arms OE and OF. Now in congruent triangles, opposite equal angles lie the equal sides:

$$BE = BF. \tag{1}$$

Now consider another pair of right-angled triangles: OEA and OCF. They also have equal hypotenuses and an arm:

$OA = OC$ (all points of the perpendicular passing through the mid-point of a segment are equidistant from its ends) and $OE = OF$. From the congruence of these triangles follows

$$AE = FC. \qquad (2)$$

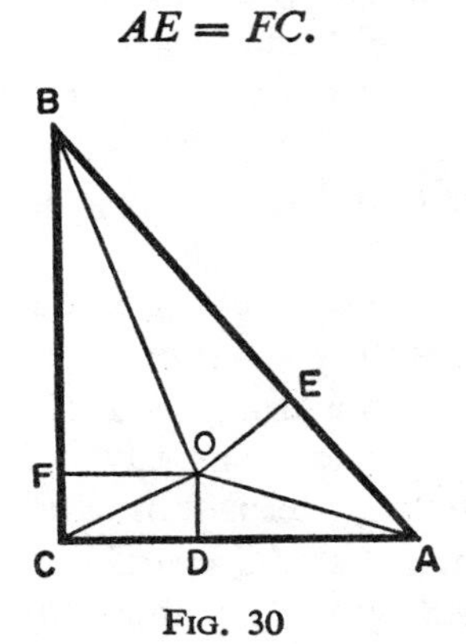

FIG. 30

Adding (1) and (2) term by term, we have

$$AB = BC.$$

60. A right-angle is equal to an obtuse angle (planar version)

Say in the quadrilateral $ABCD$ (Fig. 31) $\angle D = 90°$, $\angle C > 90°$ and $BC = DA$.

From the mid-points of the sides AB and CD construct perpendiculars which intersect at the point O. Join the point O to all the vertices of the quadrilateral.

From the pair-wise congruent right-angled triangles AKO and BKO, DLO and CLO we establish:

$$AO = OB, \quad OC = OD, \quad \angle ODL = \angle OCL. \qquad (1)$$

Since $DA = BC$, $OA = OB$, $OD = OC$, then $\triangle AOD = \triangle BOC$. From the congruence of these triangles, follow:

$$\angle ADO = \angle BCO. \qquad (2)$$

Subtracting term by term the equality (2) from equality (1), we have:

$$\angle ADC = \angle BCD,$$

i.e. a right-angle is equal to an obtuse one.

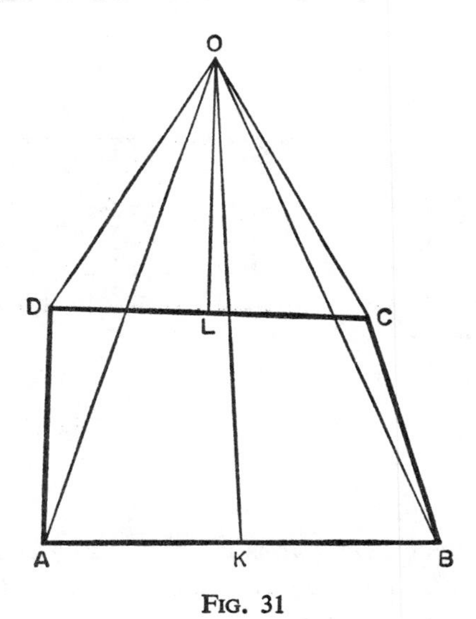

FIG. 31

61. 64 cm² = 65 cm²

Take a square of side 8 cm and cut it into four parts: two trapezoids and two right-angled triangles, as shown in Fig. 32*a*. Arranging these four parts in another order, namely in the way shown on Fig. 32*b*, we obtain a rectangle with a base of 5 cm + 8 cm = 13 cm and altitude 5 cm. The area of that rectangle is equal to 13 cm × 5 cm = 65 cm^2, while the area of the original square had been equal to 8 cm × 8 cm = 64 cm^2.

How could it happen that a simple rearrangement of the parts of the figure led to the increase of their total area? The area of the initial square is equal to 64 cm^2—as to that there can be no doubt. The area of each trapezoid obtained after the cutting is equal to $\frac{1}{2} \times (5\text{ cm} + 3\text{ cm}) \times 5\text{ cm} = 20\text{ cm}^2$, the area of each right triangle is equal to $\frac{1}{2} \times 3\text{ cm} \times 8\text{ cm} = 12\text{ cm}^2$.

Consequently, the total area of all the four parts is equal to $(20\text{ cm}^2 + 12\text{ cm}^2) \times 2 = 64\text{ cm}^2$, as it should be. From these four parts having a total area of 64 cm^2 it is in no way possible

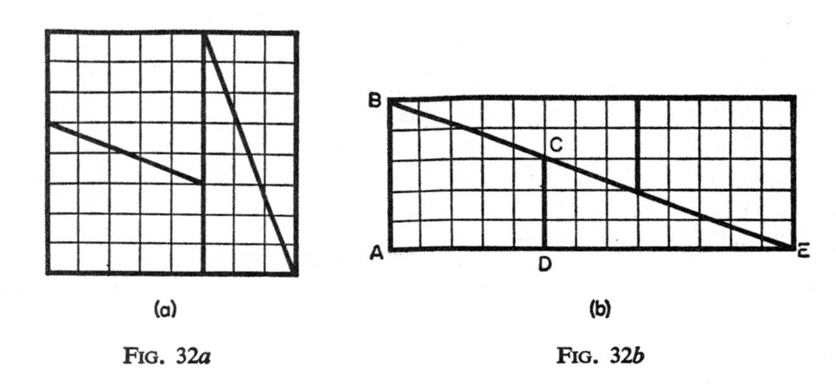

FIG. 32*a* FIG. 32*b*

to construct a figure with an area of 65 cm^2. Nevertheless we have obtained a rectangle with sides 5 cm and 13 cm, whose area is equal to 65 cm^2. The question naturally arises: is the figure obtained by us after the rearrangement really a rectangle?

62. A problem of patching

A tailor was commissioned to put a patch on a coat made of a precious fur, in order to mend a hole having the form of a scalene triangle. The tailor took the skin given to him for the patching, put it on the table fur downward, and on top of it put the coat, fur upward. Having drawn the shape of the hole, he cut out the patch and proceeded to sew it on. Then he observed that he had cut out the patch on the incorrect side. Considering the situation beyond repair, he was ready to throw out the patch and fetch material for a new one. Was he right, or was there a possibility of correcting his mistake?

II. Analysis of the Examples

43. An error was introduced in the argument: the term-by-term division of the equality (2) by the difference $AE \times DE - BE \times CE$ is inadmissible, since that difference is equal to zero, by virtue of equalities (1). Both members of equality (2) are also equal to zero.

44. In the argument under consideration, containing a considerable number of entirely correct considerations, an error is introduced in passing from the correct equality (4) to the incorrect equality (5). In fact, the difference $AE \times FH - AC \times HD$, by which we divided both members of equality (4), is equal to zero. This is clear from the proportion $AE : AC = DH : FH$, which holds for the sides of the similar triangles ACE and HFD.

45. The mistake is introduced in passing from the correct equality (5) to the incorrect equality (6). In fact, the difference

$$S'ah_1a''h''_1 - S''a'h'_1bh_2,$$

by which we had divided both members of equality (5), is equal to zero. This is clear from the analysis of the minuend and the subtrahend, each of which is equal to the product $4SS'S''$, since $\frac{1}{2}ah_1 = S$,

$$\tfrac{1}{2}a''h''_1 = S'', \quad \tfrac{1}{2}a'h'_1 = S', \quad \tfrac{1}{2}bh_2 = S.$$

The discovery of the typical error (the extension to an exceptional case) in this sophism, somewhat cumbersome in notation, requires a certain amount of independent thinking by the student.

46. The absurdity of the conclusion is obvious, but the most attentive check of the whole argument does not show up the error. We come to suspect the correctness of the reference to the property of a series of equal ratios. Let us recall how this property is proved.

Let $a_1 : b_1 = a_2 : b_2 = q$. Here b_1, b_2, q are quite arbitrary numbers, where $b_1 \neq 0$, $b_2 \neq 0$. The numbers a_1 and a_2 are defined by the equalities $a_1 = b_1 \times q$, $a_2 = b_2 \times q$. Adding these two equalities term by term we shall obtain a new equality:

$$a_1 + a_2 = (b_1 + b_2) \times q.$$

If $b_1 + b_2 \neq 0$, then this latter equality may be transformed into the form $(a_1 + a_2) : (b_1 + b_2) = q$, and we arrive at the property of a series of equal ratios:

$$a_1 : b_1 = a_2 : b_2 = (a_1 + a_2) : (b_1 + b_2).$$

As we see, this property has been proved with two essential stipulations: none of the numbers b_1 and b_2 is equal to zero, and, besides, the sum $b_1 + b_2$ is also not equal to zero. If the sum $b_1 + b_2$ is equal to zero then, by virtue of equality $a_1 + a_2 = (b_1 + b_2) \times q$ both the sum $a_1 + a_2$ is equal to zero, and the last ratio is of the form $0 : 0$. Now, as is well known, the number $0 : 0$ expresses any arbitrary number.

Thus, in applying the property of a series of equal ratios, we should always make sure that the sum of the subsequent terms is not equal to zero. If it is equal to zero, one is not permitted to apply the property of a series of equal ratios.

Was not the absurdity in the above argument about the sides of the trapezoid obtained by virtue of the fact that the application of a property of a set of equal ratios was unlawful here? Is not the sum of the second terms z and $-x$ which we had there equal to zero? In other words, are not the two segments z and x equal? As it happens, the figure confirms this suspicion. The following simple calculation shows that it is indeed so.

The two proportions obtained above from the similarity of the triangles, after being freed of the denominators, assume the following form:

$$bx - az = -by,$$
$$ax - bz = by.$$

Solving these two equations with respect to x and z, we obtain, assuming $a \neq b$,

$$x = by : (a - b), \quad z = by : (a - b).$$

In other words, the segments x and z are indeed equal.

The case $a = b$ should be considered separately. In that case the proportions yield $x + y = z$ and $x = y + z$, whence $x = y + x + y$, $2y = 0$, $y = 0$ and $x = z$. Thus, in that case also the segments x and z are equal.

The reason for obtaining the absurd conclusion is entirely clarified: having obtained the proportion $a : b = (z + y) : z$ and the ratio $(x - z) : (z - x)$, we can in no way consider that the last ratio is equal to -1, since on the strength of the equality $z = x$ it is equal not to -1 but to any arbitrary number.

47. The mistake is introduced in passing from equality (1) to inequality (2). The point is, that both terms of the first ratio of equality (1) are negative and, therefore, particular caution is necessary in comparing their values: having the proportion $(-1) : (-2) = 1 : 2$, we may not assert that $-1 < -2$.

48. I. The answer $2\text{ cm} \times 3\text{ cm} \times 10{,}000 = 60{,}000\text{ cm}^2 = 6\text{ m}^2$ is untrue, since the areas of similar figures (the rectangle $ABCD$ on the plan and the rectangle $A_1B_1C_1D_1$ in nature are similar) are not proportional to the sides, but are proportional to the square of the sides. The sides of rectangle $A_1B_1C_1D_1$ are greater than the corresponding sides of the rectangle $ABCD$ by 10,000 times, while the area is greater not by 10,000 times but $10{,}000^2$ times, and therefore the area

$$A_1B_1C_1D_1 = 6 \times 100{,}000{,}000 = 600{,}000{,}000\ (\text{cm}^2),\text{ i.e. 6 hectares.}$$

II. This assertion would have been correct if the volume V of the cylinder were proportional to the altitude h and to the base diameter d. But formula $V = 0{\cdot}25\pi d^2h$ shows that the volume of the cylinder is proportional not to the base diameter but to its square. Hence we conclude that the second drum has a volume twice as large as the first, and the third—twice as small as the first. This is confirmed by the formulae:

for the first drum $V_1 = 0{\cdot}25\pi d^2h$,

for the second drum

$$V_2 = 0{\cdot}25\pi \times (2d)^2 \times 0{\cdot}5h = 0{\cdot}5\pi d^2h = 2V_1,$$

for the third drum

$$V_3 = 0{\cdot}25\pi \times (0{\cdot}5d)^2 \times 2h = 0{\cdot}125\pi d^2h = 0{\cdot}5V_1.$$

III. The model of the structure and the structure itself are geometrically similar bodies, and the volumes of similar bodies, as is known from geometry, are proportional to the cubes of their linear dimensions, while the weights are proportional to the volumes (if the bodies are made of the same material). Therefore the increase of the dimensions of the model to the dimensions of the structure itself, namely an increase of 20-fold, will cause an increase of the volume and weight of $20^3 = 8000$ times, and the structure itself will weigh 3 kg $\times$ 8000 = 24,000 kg = 24 metric tons.

IV. A simple analysis of a few figures, be it of a simple rectangular shape, shows immediately that there exists no definite dependence between the areas and perimeters. Thus, a rectangle with sides 10 m and 15 m has a perimeter of 50 m and an area of 150 m^2, while a rectangle with sides of 5 m and 20 m having the same perimeter, has an area considerably smaller (only 100 m^2), whereas a rectangle with the sides 12 m and 13 m, has again the same perimeter of 50 m, but an area of 156 m^2. We can form judgements as to the area of two figures from their perimeters only in the case when the figures are geometrically similar. Then their areas are proportional to the squares of their perimeters. Two geometrically similar plots, having perimeters of 50 m and 60 m, have areas whose ratio is $60^2 : 50^2 = 1{\cdot}44$, and therefore the area of the first plot is greater than the area of the second by 44 per cent of the area of the latter. But if the plots with the perimeters of 60 m and 50 m are not similar, then nothing may be said as to their areas.

49. We had assumed that the larger wheel rolls along the straight line DE without sliding. This means that with every complete revolution of the wheel its centre traverses a path equal to the length of its circumference. Of course we are correct in considering that the smaller wheel set upon a common axis with the larger, and rigidly attached to it, rolls along the straight line *FG*. But will this smaller wheel roll without sliding? Of course not, since with its every revolution its centre traverses a path CC_1, not equal to the length of its circumference $2\pi r$, but one greater than that length ($CC_1 = 2\pi R$, $R > r$, $2\pi R > 2\pi r$). Consequently, the smaller wheel not only rolls along the straight line *FG*, but also slides along it.

As to the mathematical essence of that argument, it consists in the possibility of establishing a one-to-one correspondence between the set of points of any two concentric circles, apparently realized by the drawing of radii from their common centre. Of course, by the establishment of such a correspondence, one asserts not the equality of the lengths, but only the equivalence of two sets of points. Thus, we have met with a characteristic property of infinite sets, which allows one to map the whole upon its proper part [subset].

The argument under analysis is the famous so-called "Aristotelian pendulum." This problem, accounted by Aristotle as one of the most amazing in the domain of mechanics, is set forth by him as Chapter 25 of the *Problems of Mechanics*.

Aristotle's explanation of this argument was not sufficiently precise. Not satisfied with it, Galileo (1564–1642) in his *Dialogues of Mechanics* proposed his approximate solution of the problem, based on the comparison of a rolling circle with a multi-sided regular polygon successively adhering to a straight line.*

Finally, the mathematical essence of the "Aristotelian pendulum" is attained by the application of Cantor's definition of equivalent sets (G. Cantor, 1845–1918).

50. It is very easy to find the error which led to this absurd conclusion, if one recalls the exact definition of a limit: a number

* Galileo Galilei, *On Mechanics*, Univ. of Wisconsin Press, 1960.

a is the limit of a variable quantity x, if the absolute value of the difference $x - a$ tends to zero while the quantity x varies, i.e. from some instant on, it becomes and subsequently remains smaller than any previously given number. In our argument we are dealing with a constant quantity S_n, expressing the sum of all the links of the broken line BC and taking the values S_2, S_4, S_8, S_{16}, and so on. All of these values are equal to each other, since always $S_n = AB + AC$. Consequently, the quantity S_n is only apparently variable, while in effect that quantity is *constant*. But this still does not prevent us from considering the limit of S_n, as $n \to \infty$. It is clear that the limit of S_n can be no other number than the one expressing the sum of the lengths of both arms $AB + AC$; if a differs from $AB + AC$, then the absolute value of the difference $S_n - a = AB + AC - a$, being a constant quantity unequal to zero, can in no way tend to zero. Consequently, the equality (1) written above is incorrect and should be replaced by another, namely $\lim S_n = AB + AC$, i.e. by equality (2). The absurd conclusion, as we see, is eliminated.

The present sophism is very instructive. It shows how careful and critical one should be toward one's intuitive conception of a problem. The steplike broken line composed of n identical steps, which we see on Fig. 21, differs less and less from the hypotenuse BC with an unbounded increase of the number n. For sufficiently large values of n our eye is incapable of distinguishing this broken line from a straight line: imagine, for example, that we had taken n equal to 1,000,000! This broken line is connected with a whole set of variable quantities—it is possible to speak of the length of the broken line, of the area of the figure bounded by the arms AB and AC and that broken line, of the sum of the areas of all the triangles situated between the hypotenuse BC and the broken line, of the sum of the perimeters of all these triangles, of the sum of their interior angles and so on. To what limit each of these variable quantities tends, and whether it tends to any limit at all—these questions have to be solved not by intuitive representation, but by the precise definition of the concept of limit.

The intuitive picture helps (but not always even that) to make a correct guess; but the problem is solved by reasoning.

It is interesting to note that in solving the problem of the limit of the area Q_n of the figure bounded by the arms AB and AC and the steplike broken line, the intuitive picture leads to the quite correct conclusion that this limit is the area of the triangle ABC. Indeed, setting $AB = b$, $AC = h$, we find that the area of each triangle formed by two neighbouring links of the broken line and the hypotenuse is equal to $\frac{1}{2} \times \frac{b}{n} \times \frac{h}{n} = \frac{bh}{2n^2}$, and the sum of the areas of all such triangles is equal to $\frac{bh}{2n^2} \times n = \frac{bh}{2n}$. For the area of the entire figure, we obtain the formula

$$Q_n = \tfrac{1}{2}bh + \frac{bh}{2n} = \tfrac{1}{2}bh\left(1 + \frac{1}{n}\right)$$

and then the limit Q_n is $\frac{1}{2}bh$, since the limit $1/n$ is equal to 0, as $n \to \infty$.

51. The equality (2) is correct, the equality (1) is erroneous. Here, the same mistake is introduced as in Problem 50.

52. Of course, it is the formula $P = \pi rl$ that is correct and not the formula $P = \pi r(h + r)$ introduced by us. In deriving the latter we had accepted without proof that the limit of $P_n = P$, and this is untrue. Noting that $\pi r(h + r) = \pi r^2 + 2\pi \times (\frac{1}{2}r) \times h$, we find that the limit of P_n is not the side surface of the cone, but the sum of the areas of the cone base and the side area of a cylinder with a base radius $\frac{1}{2}r$ and altitude h.

53. The diagram representing the straight lines AK and AC as separate is erroneous—these straight lines coincide. In the contrary case the sum of the angles lying on one side of the straight line would exceed 180°. In fact, $\angle BAK + \angle KAC + \angle CAN = 180° + \angle KAC > 180°$. By the same method of *reductio ad absurdum* we establish that the angle KAC is zero, i.e. the straight lines AK and AC coincide.

Thus, this sophism is based on an error of construction: coincident points Z and C are considered as different.

54. This sophism is constructed on an error in the diagram: the coincident points E and F are considered as distinct. The straight line LE must satisfy two requirements: (1) it is perpendicular to PQ (since according to the construction it is perpendicular to MN, and $MN \parallel PQ$), i.e. it forms a right-angle with that line, and the sides of the right-angle pass through the ends of the diameter KL; (2) it is a side of the inscribed angle standing upon the diameter KL.

From the above we conclude that the vertex of the inscribed angle must be both on the straight line PQ and on the semicircle, i.e. at the point of their intersection F.

55. The two assumptions made about the position of the point B relative to the circle do not exhaust all the possibilities: this point B may be not inside and not outside but upon the circumference.

Then the two points G and H of intersection of the circle with the sides of the angle become coincident at one point, coinciding with B, and instead of two diameters GF and HF we obtain one, namely BF. The contradiction to which the assumption that the point B is located outside, or inside, the circle leads, proves that the only correct assumption is the third, for which the incorrect conclusion disappears.

56. The diagram is the cause of the contradiction. By denoting the second point of intersection of the circles I and II by the letter F and connecting F with the points A, B, and C, it is easy to see that $\angle AFB = \angle CFB = 90°$ and therefore that the segments AF and CF lie on a single straight line AC. Consequently, this straight line AC is intersected by the circles I and II not at two different points D and E but at a single point F, and there exists only one perpendicular BF, dropped from the point B to the straight line AC.

57. Of course, the argument shows only that the point D lies on the straight line BC, coinciding with the point E—the circles constructed on the sides AB and AC of the triangle ABC as diameters, intersect at the point E, which is the base of the perpendicular dropped from the vertex A to the side BC.

58. An accurate construction of the diagram will show immediately the reason for the contradiction, but it may be established also without a new diagram by the following reasoning.

About $\triangle ABC$ circumscribe a circle. The axis of symmetry DD_1 which is perpendicular to the chord AB passes through the mid-point of the arc AB subtended by that chord. But the bisector of the angle ACB also has to pass through the mid-point of that arc—otherwise the inscribed angles ACC_1 and BCC_1, measured by one-half of the corresponding arcs, would not be equal to each other. Consequently, this mid-point of the arc AB, being the common point of the bisector and the median, is precisely their point of intersection E. Thus, the point E is invariably outside the triangle—the first and third assumptions drop out, and there remains the second assumption, to which Fig. 29*b* corresponds.

Thus, the quadrilateral $AEBC$ turns out to be inscribed in a circle. But in every quadrilateral inscribed in a circle the sum of every pair of opposite angles is equal, as is well known, to 180°, and therefore two assumptions may be made about the angles EAC and EBC—either both of them are right-angles, or one of them is acute and the other obtuse. If the first is the case, then the perpendiculars EF and EG, dropped to the sides AC and BC, coincide with the sides EA and EB, and the right-angled triangles ACE and BCE are equal, as having a common hypotenuse CE and equal arms $AE = BE$. Consequently, $AC = BC$, which contradicts the condition $AC > BC$. Thus, the assumption that $\angle EAC = \angle EBC = 90°$ is rejected, and of these two angles one is acute, and the other is obtuse. Hence we conclude that the position of the triangles AEF and BEG, shown in Fig. 29*b*, does not correspond to reality—both points F and G cannot be outside the triangle ABC, since then both angles EAC and EBC would be obtuse. The only correct disposition is the one shown in Fig. 33, where one of the points F and G is inside the triangle ABC, and the other is outside it. But then from the equalities $CF = CG$ and $AF = BG$ it does not follow that $AC = BC$, since $AC = AF + CF$ and $BC = CG - BG$, and no contradiction with the condition $AC > BC$ follows.

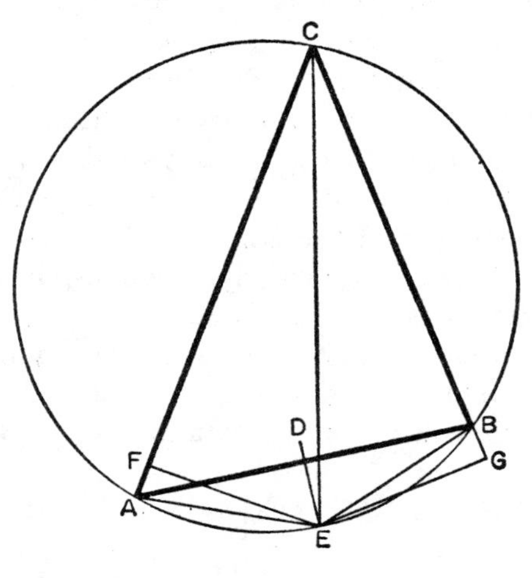

FIG. 33

59. In the "proof" not all the possible assumptions have been considered. In fact, we have limited ourselves to the assumption that the bisector of the angle B and the axis of symmetry of the segment CA (i.e. the straight line perpendicular to CA at the mid-point E of that segment) intersect within the triangle ABC. Meanwhile, we should have considered also all the other possible assumptions: (1) that the point of intersection lies on the segment CA, (2) that the point of intersection is outside the triangle ABC.

By analysing all the cases it would have appeared which of them were possible.

We shall demonstrate that the only possible case is the third one—in any right-angled triangle ABC the bisector of the angle

B intersects the axis of symmetry of the arm *CA* outside the triangle.

About $\triangle ABC$ circumscribe a circle (Fig. 34). Its centre being on the mid-point of the segment *AB*, we shall denote it by the letter *K*. It is easily seen that the radius *KO*, perpendicular to the

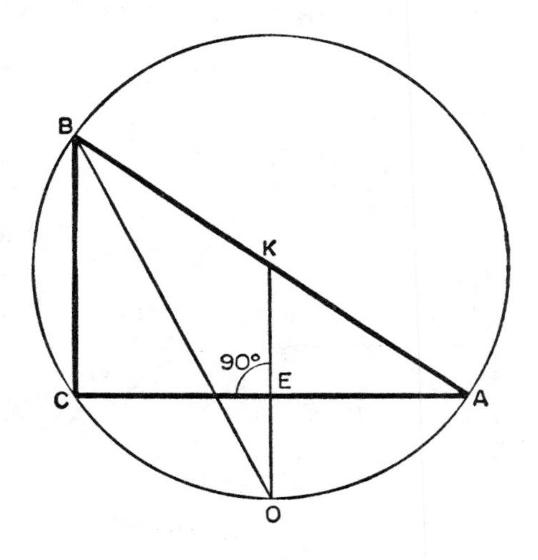

FIG. 34

chord *CA*, will bisect this chord and also the arc subtended by it. Thus, the point *O* is the mid-point of the arc *CA*.

We now draw the bisector of the angle *B*. Since the point *B* lies on the circumference and the bisector divides the angle *B* in halves, then the bisector of the angle *B* also passes through the point *O*.

Consequently, the intersection of the bisector of the angle *B* with the axis of symmetry of the arm *CA* in any right-angled triangle *ABC* lies outside that triangle.

This eliminates the possibilities of the absurd conclusion.

60. In the "proof" not all possible assumptions have been considered. In fact, besides the one considered, there were also the following: (1) the point O is inside the quadrilateral $ABCD$; (2) the point O lies on DC, i.e. is the mid-point of that segment.

However, in those cases also it is easy to obtain the same absurd conclusion: a right-angle is equal to an obtuse one.

To clarify the misunderstanding, it should be noted that the case considered in the text should, in its turn, be subdivided into

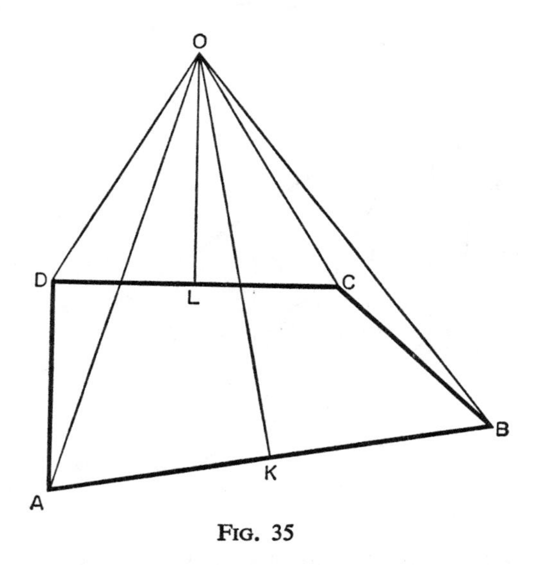

FIG. 35

two cases, namely the obtuse angle BCD and the triangle BOC lie (1) to one side of the straight line BC (Fig. 31); (2) on different sides of the straight line BC (Fig. 35).

The first assumption leads to an absurd conclusion, of which we have already convinced ourselves.

The second assumption does not lead to an absurdity—the right angle ADC, just as before, is represented by the difference of two angles ($\angle ADO$ and $\angle ODC$), and the obtuse angle BCD supplements up to 360° the sum of two such angles ($\angle BCO$ and $\angle OCD$). Arguing by the method of *reductio ad absurdum* we establish that the second assumption of the third case is the only possible one.

61. The trapezoid $ABCD$ with the bases $AB = 5$ cm and $CD = 3$ cm and altitude $AD = 5$ cm has right-angles at the vertices A and D. Placing next to it triangle CDE with a right-angle at the vertex D and the arm $CD = 3$ cm and $DE = 8$ cm, we obtain a new figure bounded below by the rectilinear segment AE, since two right-angles ADC and CDE together give a straight angle. But will the sides BC and CE lie on a single straight line? Even though it seems so with the naked eye, a carefully drawn diagram clearly shows that in reality the line BCE is a broken and not a straight one. Every doubt on that point is removed if we note that when BCE is a straight line, we obtain a pair of similar triangles ABE and CDE, in which the ratio of the vertical arms AB and DC, lying opposite the common angle E, is equal to $5 : 3 = 1{\cdot}666\ldots$ while the ratio of the horizontal arms AR and DE is equal to $13 : 8 = 1{\cdot}625$. Thus the figure obtained after the rearrangement is not a rectangle. It is transformed into a rectangle only if we add to it a long and narrow parallelogram, clearly shown in Fig. 36, and having an area of exactly 1 cm^2.

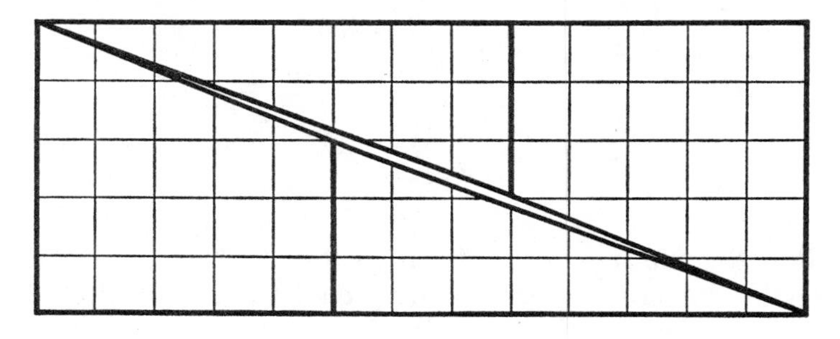

FIG. 36

Thus, in the present misunderstanding it is our eye which is guilty of not noticing the small difference in the directions of the segments BC and CE.

The conclusion obtained from the special case $(64 = 65)$ admits of a generalization.

We consider three diagrams representing figures composed of the same pieces, namely: two congruent trapezoids and two congruent triangles (Fig. 37*a*, 37*b*, 37*c*).

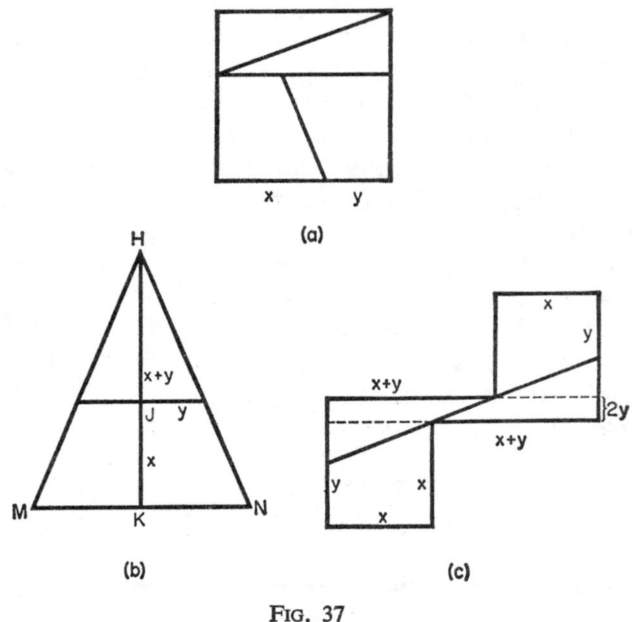

FIG. 37

Denoting the area of each figure by the letter S with the corresponding subscript, we write down the following equalities:

$$S_1 = (x + y)^2 = x^2 + 2xy + y^2 \text{ (Fig. 37}a\text{)};$$

$$S_2 = \tfrac{1}{2}MN \times HK = \tfrac{1}{2} \times 2x \times (2x + y)$$

$$= 2x^2 + xy \text{ (Fig. 37}b\text{)};$$

$$S_3 = (2x + y) \times (2y - x) + 2x^2$$

$$= 4xy - 2x^2 + 2y^2 - xy + 2x^2 = 3xy + 2y^2 \text{ (Fig. 37}c\text{)}.$$

Let us establish the condition under which the equality

$$S_1 = S_2 = S_3. \tag{1}$$

will hold. For this we find the differences:

$$S_2 - S_1 = x^2 - xy - y^2,$$
$$S_3 - S_1 = y^2 + xy - x^2 = -(x^2 - xy - y^2).$$

Thus, the relation (1) holds when the equality

$$x^2 - xy - y^2 = 0 \tag{2}$$

is satisfied. We solve that equation with respect to x:

$$x = \frac{y}{2} \pm \sqrt{\left(\frac{y^2}{4} + y^2\right)} = \frac{y(1 \pm \sqrt{(5)})}{2}$$

Carrying out the choice of sign, we shall determine the ratio of x to y:

$$\frac{x}{y} = \frac{1 + \sqrt{(5)}}{2}. \tag{3}$$

Ratio (3) shows that x and y are incommensurable segments, i.e. their ratio cannot be expressed by a rational number. Hence we arrive at the idea that, if we take the numbers x and y as rational, in particular as positive integers, but such that their ratio is sufficiently close to the value of the fraction $\frac{1}{2}(1 + \sqrt{5})$, then, by transposing the parts of the figure, the difference in direction taken by the segments bounding the parts of the figure will remain outside our capacity to notice.

First of all we observe a property of the function $F(x, y) = x^2 - xy - y^2$, that $F(x, y) = -F(x + y, x)$. Indeed,

$$\begin{aligned} F(x + y, x) &= (x + y)^2 - (x - y)x - x^2 \\ &= x^2 + 2xy + y^2 - x^2 - xy - x^2 \\ &= -(x^2 - xy - y^2) = -F(x, y). \end{aligned}$$

The established pattern allows us to indicate successive integral values of the argument which do not change the absolute value of the function.

i	x_i	y_i	$F(x_i, y_i)$	$\frac{x_i}{y_i}$	*Maximum absolute error*
1	0	1	-1	0	—
2	1	0	1	—	—
3	1	1	-1	1	0·619
4	2	1	1	2	0·382
5	3	2	-1	1·5	0·119
6	5	3	1	$\frac{5}{3} \approx 1{\cdot}667$	0·049
7	8	5	-1	1·6	0·019
8	13	8	1	1·625	0·007
9	21	13	-1	$\frac{21}{13} \approx 1{\cdot}615$	0·003

The table contains the successive pairs of integral non-negative values of x and y for which the absolute values of the differences $S_2 - S_1$ and $S_3 - S_1$ are equal to unity.

Its law of construction is very simple:

$$y_{i+1} = x_i, \quad x_{i+1} = x_i + y_i, \quad \text{where } i \text{ equals } 1, 2, 3, \ldots$$

To obtain the complete illusion in rearranging the figures one should take the values from the table beginning with the sixth row. Taking 5 and 3 for x and y respectively, we obtain an approximation to the irrational fraction $\frac{1}{2}(\sqrt{5} + 1)$, which differs only in the hundredths of unity, while taking 21 and 13—in the thousandths, and so on, with an ever growing degree of accuracy.

62. If we take a scalene triangle ABC (Fig. 38a) and invert it to the other side, then after making the point A coincide with the point C and C with A, it will take the position of the triangle AB_1C, not coinciding with the triangle ABC. But is it not possible to cut the triangle ABC into such portions that each part separately admits of an inversion? Clearly, every isosceles triangle admits of an inversion, and therefore the problem is: to cut up the given scalene triangle in such a way, that every portion represents an isosceles triangle. Figure 38*b* shows how this is to be done. We first draw $BD \perp AC$, then from the vertex D of the right-angle of each of the right-angled triangles so obtained, we draw the medians DE and DF (the point E is the mid-point of the segment AB, and the point F is the mid-point of the segment BC).

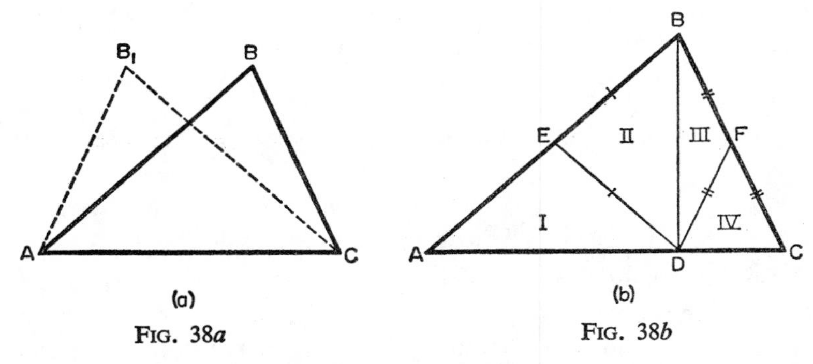

FIG. 38*a*

FIG. 38*b*

It is easy to show that the median of every right-angled triangle from the vertex of the right-angle is equal to one-half the hypotenuse. Therefore, all the four triangles ADE, BDE, BDF, CDF obtained on Fig. 38*b* are isosceles. ($AE = DE = BE$, $DF = CF = BF$). Cutting the prepared patch into four such parts, and then inverting each of them separately, the tailor will be able to correct his mistake (of course, he will not only have to sew the whole patch into place, but also first to sew together the first, second, third and fourth triangles). In order to clarify the whole matter, we recommend the reader to cut out the Fig. 38*b* from paper, one side of which is coloured.

Note that an isosceles triangle is not by any means the only figure admitting of an inversion. An inversion is admitted by every figure having at least one *axis of symmetry*. Among such figures are every rectangle (2 axes of symmetry), and square (4 axes of symmetry), and rhombus (2 axes of symmetry), and every regular n-gon (n axes of symmetry), and a circle (every diameter serves as the axis of symmetry), and an infinite set of more complicated figures.

Based on the above, we simplify the solution of the problem of the patch. In fact, the rectangle $BFDE$, having the axis of symmetry EF, admits of an inversion. Consequently, in order to correct the tailor's mistake, it is sufficient to make not three but two cuts, carrying them out in the directions DE and DF. This is, of course, more expedient, since one will have to sew together not four but three pieces.

III. Stories, with Explanations, of Causes of Erroneous Reasoning

63. Similar triangles with equal sides

Take two similar scalene triangles and denote the sides of the first in the order of increasing size by the letters a, b, c ($a < b < c$), and the corresponding sides of the second by the letters a_1, b_1, c_1. By virtue of proportionality of the corresponding sides of similar polygons we have: $a_1 = aq$, $b_1 = bq$, $c_1 = cq$, where q is the coefficient of proportionality, and therefore $a_1 < b_1 < c_1$.

If $q = 1$, then all the sides of the two triangles are respectively equal, and the triangles are congruent. The congruence of the triangles is thus a special case of similarity.

It may seem, that if the triangles are similar, but not congruent, then they have no equal sides. The erroneousness of such a conclusion is shown by a simple consideration of triangles with the sides 8, 12, 18 cm, and 12, 18, 27 cm. The sides of the second are one-and-one-half times as large as the corresponding sides of the first, and therefore the triangles are similar (but not congruent). As we see, these two triangles have two pairs of respectively equal sides.

We shall establish the conditions which have to be satisfied by two similar, but not congruent, triangles having two pairs of respectively equal sides.

Suppose that $q > 1$. The smallest side a of the first triangle (with the sides a, b, c) is less than the smallest side a_1 of the second triangle (with sides $a_1 = aq$, $b_1 = bq$, $c_1 = cq$), and cannot be equal to any of the sides of the latter.

The middle side b of the first triangle may be equal only to the smallest side a_1 of the second triangle (since b is less than $b_1 = bq$ and much smaller than $c_1 = cq$), and the greatest side c of the first triangle is equal either to the smallest side a_1 or to the middle side b_1 of the second triangle. If two sides of the first triangle are equal to two sides of the second triangle, then $b = a_1$, $c = b_1$. Hence $b = aq$, $c = bq = aq \times q = aq^2$. Thus the sides of the first triangle form a geometric progression a, aq, aq^2. The sides of the second triangle are equal to aq, aq^2, aq^3, and represent the second, third, and fourth terms of that same geometric progression.

The side a may be arbitrary. But the number q is not completely arbitrary. It has to satisfy the inequality $aq^2 < a + aq$, since the greatest side of the triangle must be less than the sum of the two other sides. Dividing both members of that inequality by a, we obtain the inequality $q^2 - q - 1 < 0$, which has to be satisfied by the number q (consideration of the second triangle leads to the same inequality). By factoring the second degree trinomial $q^2 - q - 1$, we rewrite this inequality in the form

$$\{q - \tfrac{1}{2}(1 + \sqrt{5})\}\{q - \tfrac{1}{2}(1 - \sqrt{5})\} < 0$$

or in the form $(q - 0{\cdot}5 - 0{\cdot}5\sqrt{5})(q - 0{\cdot}5 + 0{\cdot}5\sqrt{5}) < 0$, and we note that the expression in the second bracket is always positive for $q > 1$. Consequently, the expression in the first bracket must be negative, and therefore:

$$q < 0{\cdot}5 + 0{\cdot}5\sqrt{(5)},\ q < 0{\cdot}5 + 0{\cdot}5 \times 2{\cdot}236\ldots,\ q < 1{\cdot}618\ldots$$

Thus, two similar unequal triangles have two pairs of respectively equal sides if, and only if, the sides of the first triangle are equal to a, aq, aq^2, and the sides of the second are equal to aq, aq^2, aq^3, where the side a is arbitrary and the number q is taken between 1 and $0{\cdot}5 + 0{\cdot}5\sqrt{(5)} \simeq 1{\cdot}618$.

We can find any number of pairs of such triangles.

64. Trisection of an angle

To carry out the trisection of an angle means to divide an angle into three equal parts. To do this is, of course, not at all difficult. We may, for example, measure the given angle with a protractor, divide the measured number of degrees by three, and then, by means of the same protractor, lay off an angle containing the quotient in degrees. We can, however, manage also without the protractor, applying the method of "successive approximation" —having constructed an arc of arbitrary radius, for which the given angle is the central angle, we lay off by eye a chord corresponding to a third part of the arc, and lay off that chord successively three times along the arc, beginning from one of its ends. If we then find ourselves on the other end of the arc, the problem is solved. If, however, as usually happens, we do not reach the other end of the arc, or pass beyond it, then the chord taken by eye has to be corrected, increasing or decreasing it by one-third of the distance from the point obtained to the end of the arc, where this one-third we take again by eye. This corrected chord is again laid off on the arc, and in case of necessity we correct it again in the same manner. Each new (corrected) chord will give an ever more exact solution and, finally, after repeating the operation a few times, we obtain a chord which will be laid off the given arc almost exactly three times, and the trisection of the angle will have been carried out. Obviously, these two methods allow us to divide the given angle not only into three but into any number of equal parts.

However, when mathematicians speak of the problem of the trisection of an angle, they have in view not these methods, worthwhile from a practical point of view yet only approximate, but an exact method, based exclusively on the application of a ruler and compass. It is also necessary to note that only one arm of the ruler may be used and that the ruler should serve only for drawing straight lines (the use of scaled rulers is not allowed), and the compass may be used only for drawing arcs. Finally, the method should yield the solution of the problem by a finite number of operations of drawing straight lines and arcs of circles. The last remark is very essential. Thus, having established (from the formula for the sum of an infinitely decreasing geometric progression) that

$$\frac{1}{3} = \frac{1}{4} + \frac{1}{16} + \frac{1}{64} + \frac{1}{256} + \ldots,$$

it is possible to propose the following solution for the problem of the trisection of an angle, requiring the application of only a ruler and compasses—divide the given angle into four equal parts which, as is well known, may be carried out by means of the compass and ruler, and then to the angle obtained add a correction, equal to one-fourth of it, i.e. $\frac{1}{16}$ of the given angle, then a second correction, equal to $\frac{1}{4}$ of the first, i.e. $\frac{1}{64}$ of the given angle, and so on. The exact solution of the problem by this method requires an infinitely great number of operations (divisions of an angle into four equal parts) and therefore is not that classical solution which one has in view when speaking of the solution of the problem of the trisection of an angle and other problems of construction.

Thus, we shall be dealing with the exact solution of the problem of trisection of an angle by means of carrying out a finite number of straight lines and circles.

For some angles this problem is very easy to solve. Thus, to trisect an angle of 180° it is sufficient to construct an angle of 60°, i.e. the angle of an equilateral triangle, and to trisect angles of 90 and 45°—angles of 30 and 15°, i.e. half and one-quarter of the angle of an equilateral triangle. However, it has been proved that, besides the infinite set of angles admitting of trisection, there exists also an infinite set of angles not admitting of trisection (in the sense indicated above). Thus, it is impossible to divide into three equal parts (by construction of a finite number of lines and circles) an angle of 60°, or an angle of 30°, or an angle of 15°, or an angle of 40°, or an angle of 120°, or an infinite set of other angles.*

We shall now consider whether the following method, often recommended, for dividing an arbitrary angle *ABC* into three equal parts is correct. From the vertex *B* draw an arc of a circle of arbitrary radius, intersecting the sides of the angle at the points *D* and *E* (Fig. 39). Divide the chord *DE* into three equal parts

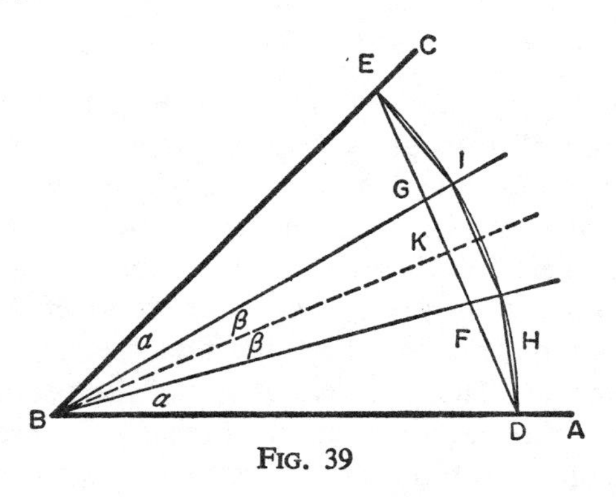

FIG. 39

and join the points of division *F* and *G* to *B*. It seems that the angles *DBF*, *FBG*, *GBE* will turn out to be equal, and then the trisection of the arbitrary angle *ABC* will consequently have been carried out as required, i.e. by drawing a finite number of straight lines and circles: the division of the segment *DE* into three equal parts, which was required here, is carried out, as is well known, in just this way.

Those who propose this solution assume that the equality of

* See, for example, the books: A. Adler, *Theory of Geometric Constructions* (Teoriya geometricheskikh postroeniy), pp. 173–180, Leningrad, 1940. I. Aleksandrov, *Problems in Geometrical Constructions and Methods of Their Solution* (Geometricheskie zadachi na postroenie i metody ikh resheniya), pp. 144–150, Moscow, 1934. B. Argunov, and M. Balk, *Geometric Constructions in the Plane* (Geometricheskie postroeniya ploskosti), pp. 214–218, Moscow, 1957.

the segments DF, FG, GE, into which we have divided the chord DE, implies also the equality of the arcs DH, HI, IE, obtained by protracting BF and BG up to the intersection with the circle. Is it so? If these arcs are equal, then also the angles DBH, HBI, IBE are equal (say each of them is equal to α); and the chords DH, HI, IE subtending them are also equal. But the segment HI is greater than the segment FG (this assertion is suggested by the diagram, but we shall prove it below), and the segment DH is equal to the segment DF, since the angles DFH and DHF are equal:

$$\angle DFH = \angle DBH + \angle BDF = \alpha + \tfrac{1}{2}(180° - 3\alpha) = 90° - \tfrac{1}{2}\alpha,$$

$$\angle DHF = \tfrac{1}{2}(180° - \alpha) = 90° - \tfrac{1}{2}\alpha.$$

Consequently, when the segments DH and HI are equal, segments DF and FG, contrary to expectation, are unequal, and the assumption of the equality of DH and HI has to be rejected.

Dropping a perpendicular BK from the vertex B to the chord DE, we note that the entire figure is symmetrical about BK: folding the diagram along BK, we bring its halves into coincidence. Hence we conclude that the segment HI is perpendicular to BK, and on the strength of this the segment FG is parallel to HI, and the triangles BHI and BFG are similar, yielding: $HI : FG = BH : BF$. But $BH > BF$, and therefore also $HI > FG$, as we asserted above.

Thus, the division of a chord into three equal parts does not yield a division of the corresponding arc into three, and consequently the trisection of the corresponding central angle is not achieved: the middle angle will invariably turn out to be somewhat greater than each of the outside ones. True, with a small angle ABC the difference will not be large, and in practice one sometimes uses this method for the approximate division of a small angle into three equal parts.

We shall give one more proof of the incorrectness of the assumption that from the equality of the segments DF, FG, GE

follows the equality of the arcs DH, HI, IE and the angles DBH, HBI, IBE. Assuming that all this is so, and setting $\alpha = 2\beta$, we easily obtain the formulas: $DK = BK \times \tan 3\beta$, $FK = BK \times \tan \beta$. But $DK = 3FK$, whence $\tan 3\beta = 3 \tan \beta$. However, applying the formula for the tangent of the sum of two angles, we easily find that

$$\tan 3\beta = \frac{3 \tan \beta - \tan^3 \beta}{1 - 3 \tan^2 \beta}.$$

Comparing this correct relation with the formula $\tan 3\beta = 3 \tan \beta$ obtained above, we arrive at the equality $9 \tan^3 \beta = \tan^3 \beta$ or $8 \tan^3 \beta = 0$, true only for $\tan \beta = 0$. Hence it follows that, with the equality of the segments DF, FG, GE, the corresponding central angles cannot be equal.

65. More about the trisection of an angle

Here is a method for dividing an arbitrary angle into three equal parts (by means of an "insert"), pointed out in ancient Greece (possibly by Archimedes, who lived *ca.* 287–212 B.C.).*

It can be shown that this method yields that solution of the problem of trisection of an arbitrary angle, about which it was said above that it is impossible for an infinite set of angles.

Extend one of the sides of the given arbitrary angle $\angle ABC = \alpha$ (Fig. 40) beyond the vertex B and draw a semicircle with arbitrary radius r and centre at B. Let that semicircle intersect the second side of the angle at the point D. Then take a straight-edge and make on it two marks E and F at a distance r apart. Place the straight-edge in such a way that its arm passes through the point D and that the point E is on the extension of BA. The mark F will then be either outside the semicircle, or inside it, or upon it. In the first two cases, move the straight-edge, observing both conditions indicated (the straight-edge to pass through the point

* Tseiten, G., *History of Mathematics in Ancient Times and the Middle Ages* (Istoriya matematiki v drevnosti i v srednie veka), pp. 64–67, Moscow, 1932.

D, and the mark E to lie on the extension of BA), till the mark F comes on the semicircle. We have, so to speak, carried out an "insertion" of the segment $EF = r$ between the straight line BA and the circle, where this "insertion" is intersected by the arc originating from the point D.

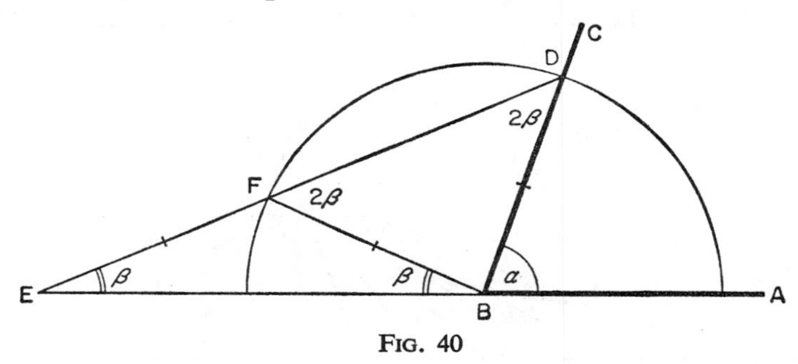

FIG. 40

Now we have obtained an isosceles triangle BFE ($EF = BF = r$). Denoting each of the two equal angles at its base by β, we have that $\angle BFD = 2\beta$. But the triangle BDF is also isosceles, and therefore $\angle BDE = \angle BDF = 2\beta$. It remains to consider the triangle BDE, for which the angle $ABC = \alpha$, as an exterior angle, is equal to the sum $\angle BED = \beta$ and $\angle BDE = 2\beta$. Thus, $\alpha = 3\beta$, and $\angle BED$ ($= \beta$) is, consequently, equal to one-third of the given arbitrary angle α.

We have solved the problem of trisection of an arbitrary angle by means of a ruler and compass, and at first sight it seems that this solution refutes what had been said before about the impossibility of trisecting very many angles. But a more careful analysis shows that there is no contradiction here whatsoever. It had been proved that not every angle may be divided into three equal parts by means of drawing a finite number of straight lines and circles. But in the solution by the method of "insertion," the ruler was used not only for drawing straight lines, but also for carrying out a more complicated operation—the insertion of a radius between the extension of BA and the semicircle—the straight-edge has been used in a way not envisaged by the theory. In effect, we have made use of the straight-edge to trace a particular curve, the so-called *conchoid*: by rotating the ruler about the point D and, at the same time, displacing it in such a way that the point E always lies on the straight line BA, we make the second mark (the point F) move on the plane, tracing a curve

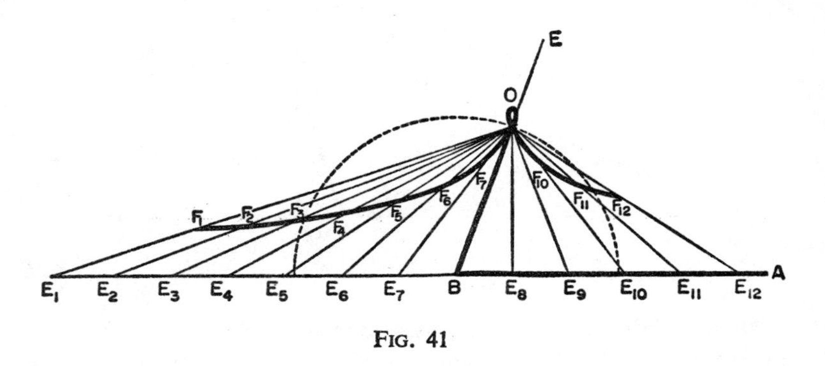

FIG. 41

shown in Fig. 41. This curve is called the conchoid. In our solution we made use of the point F_3 of intersection of the conchoid with the semicircle.

Thus, the solution of the problem of the trisection of an angle by the method of "insertion" is based on the construction, in addition to straight lines and circles, of a conchoid, and in no way does it refute the theorem, proved in the theory of geometric constructions, of the impossibility of trisecting an arbitrary angle by means of a finite number of straight lines and circles.

66. Quadrature of the circle

The famous problem of the quadrature of a circle was studied long before our time, and it was finally solved in 1882.* The problem is to construct a square having an area equal to the area

* See the history of the problem in Klein, Felix, *Famous Problems of Elementary Geometry*, Dover reprint, New York.

of a given circle. Denoting the radius of the given circle by the letter r, and the side of the required square by the letter x, we have the equation $\pi r^2 = x^2$, from which we find $x = r\sqrt{\pi}$. Since the number π, expressing the ratio of the length of a circle to its diameter, is known to a very high accuracy, and the extraction of the square root of any number may be carried out with an arbitrarily high accuracy, we easily obtain the following expression for x:

$$x \approx r\sqrt{3{\cdot}14159265} \approx r \times 1{\cdot}77245385,$$

where the values of π and $\sqrt{\pi}$ are taken to eight decimals. Nevertheless, when one speaks of the solution of the problem of the quadrature of a circle, one has in mind not the calculation, but the construction of the side of the square given the radius of the circle, such construction to be carried out by drawing a finite number of straight lines and circles (i.e. by using only the straight-edge and compass), also deriving the side of the square exactly, and not approximately. It has been shown that in such a formulation this problem is insoluble. However, attempts to solve this problem, usually by people who know little of mathematics, continue up to the present day.

We shall consider a method, which has been known for a long time, for constructing a square equal to a given circle. Take a right circular cylinder, whose base is a given circle of radius r, while the altitude of the cylinder is $\frac{1}{2}r$. If we roll this cylinder without sliding along a plane, upon which it is laid so that its axis is parallel to the plane, then in one revolution it will cover on the plane a rectangle equal to its "curved" area, and namely a rectangle with sides $2\pi r$ and $\frac{1}{2}r$. But the area of that rectangle is equal to $2\pi r \times \frac{1}{2}r = \pi r^2$, i.e. equal to the area of the given circle. Now there remains only to turn the rectangle into an equal square, and for this, as is well known, it is sufficient to construct a segment which is the mean proportional of two adjacent sides of the rectangle.

It would be a considerable error to think that this solution refutes the previous statement on the impossibility of solving the problem of the quadrature of the circle. But here, besides the compass and the straight-edge, which have been necessary for dividing the radius r in halves and for constructing the mean proportional between $2\pi r$ and $\frac{1}{2}r$, we have used one more instrument, namely the cylinder we rolled along the plane.

67. On a proof of the theorem of the sum on the interior angles of a triangle

Take an arbitrary triangle ABC and walk around it along its perimeter from the vertex A through vertices B and C and back to A. Imagine, that while carrying out this circumambulation I hold in front of me an arm, stretching it in the direction of my motion. Moving from A to B, I keep the direction of my arm unchanged. Reaching vertex B, I turn my arm counterclockwise by an angle B_1BC (Fig. 42). Further, in moving from B to C the

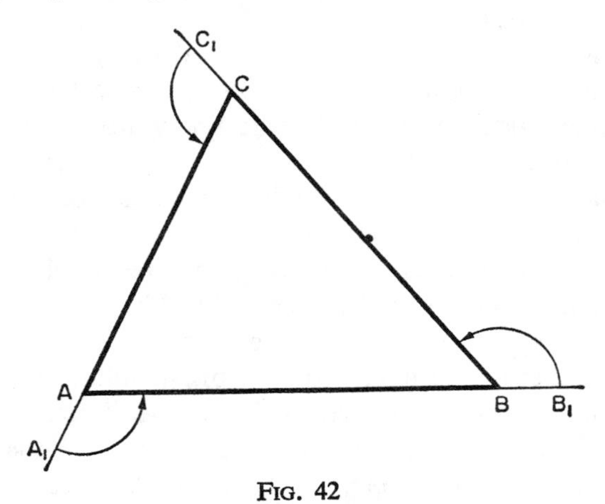

FIG. 42

direction of the arm again remains unchanged. At point C the arm makes a new turn—by an angle C_1CA. Then, in moving from C to A, the direction of the arm does not change and, finally, at A the arm makes the last turn—by an angle A_1AB.

The circumambulation finished, I have returned to the initial point, the arm has returned to its original position—it is again directed from A to B. In the course of its circumambulation, the arm has undergone one complete revolution, i.e. has rotated 360°. But this complete revolution is the sum of three rotations, namely by the angles B_1BC, C_1CA, A_1AB, which are exterior for the given triangle ABC. Thus, $\angle B_1BC + \angle C_1CA + \angle A_1AB = 360°$. But each of the exterior angles may be replaced by the difference between 180° and the corresponding interior angle. Therefore we have:

$$(180° - \angle B) + (180° - \angle C) + (180° - \angle A) = 360°,$$

where $\angle A$, $\angle B$, $\angle C$ are the interior angles of the triangle ABC. Removing the brackets and collecting similar terms, we arrive at the equality:

$$\angle A + \angle B + \angle C = 180°.$$

This simple and understandable proof is not based on any theorems of geometry, other than the theorem on the sum of two adjacent angles, does not, specifically, refer to the theorems on parallel lines, and does not depend, therefore, on the postulate on parallel lines. This state of affairs would be of tremendous advantage for our proof, if only in the course of the proof we had not based ourselves, without noticing it, upon a certain new axiom.

That this is so, we may convince ourselves as follows. Take a sphere and on it three points D, E, F, in such a way that the arcs DE, EF, FD of the great circles of the sphere are each equal to 90° (for point D one may take the North pole, with points E and F on the equator). The argument given above, proving that the sum of the interior angles of a plane triangle ABC is equal to 180°, may also be applied, without any change, to the spherical triangle DEF. We must only keep in mind that by the direction of an arc on a sphere we mean the direction of the tangent to that arc (at the point under consideration). Thus, our argument shows that also the sum of the interior angles of a spherical triangle *DEF* is equal to 180°. But this is untrue, since each of the interior angles of the spherical triangle is equal to 90°, and their sum, consequently, is equal to 270°.

In circumambulating a spherical triangle with one arm stretched along the direction of motion, and returning to the initial position, we thus sweep out by the arm a rotation not of 360°, as in the case of circumambulation of a plane triangle, but some other angle. In asserting that after the circumambulation of a plane triangle and return to the initial position we have a revolution of 360°, we are enumerating not what is self-evident and true under all conditions, but are basing ourselves on the following property of the plane, which is not possessed by the sphere—the circumambulation of every triangle on the plane involves a revolution by 360°. Taking this property of the plane as evident, we are introducing a new axiom.

Thus, the assertion that we have proved the theorem on the sum of the interior angles of a triangle without making use of either the parallel postulate, or of some other new axiom, is erroneous. Even more than a hundred years ago our genial geometer Nikolay Ivanovich Lobachevsky (1792–1856) proved that it is impossible to eliminate from (Euclidean) geometry the postulate on parallels without introducing in its place some other axiom.

68. How to compute the volume of a truncated pyramid

As is well known, the area of a trapezium may be obtained by taking the product of one-half the sum of its bases and the altitude, or by taking the product of its mean line by the altitude. Both methods yield one and the same thing, since the trapezium is equivalent to a rectangle whose base is the mean line of the trapezium, and the altitude is the same as the altitude of the trapezium.

In this connexion there arises the question—is it not possible to compute the volume of a truncated pyramid, instead of using the usual method based on the formula

$$V = \tfrac{1}{3}h(B_1 + B_2 + \sqrt{(B_1 B_2)}),$$

where h is the altitude of the truncated pyramid, B_1 and B_2 are the areas of its two bases, by another method based on the replacement of the truncated pyramid by a prism whose base is the mean base of the truncated pyramid? The mean cross-section of the truncated pyramid is the section taken parallel to its bases through the mid-point of its altitude.

Note that in technical procedures the volume of a truncated pyramid (for example, in measuring a pile of sand prepared for road work) is computed just this way: one finds the area of the mean section, and then multiplies it by the altitude of the truncated pyramid.

Many people think that this method gives completely precise results just as precise results are given by the calculation of the area of a trapezoid as the product of the mean line by the altitude. We shall clarify whether it is so by limiting ourselves to the consideration of only the simple case when the bases of the truncated pyramid are squares with sides a and $b < a$.

The volume of such a truncated pyramid is expressed by the formula $V = \frac{1}{3}h(a^2 + b^2 + ab)$. The middle section represents a square of side $\frac{1}{2}(a + b)$. The volume of the prism whose base is that mean section and whose altitude is equal to the altitude of the truncated pyramid is calculated from the formula $V_1 = \frac{1}{4}h(a + b)^2$.

Let us find the difference $V - V_1$. As a calculation shows, it is equal to $\frac{1}{12}h(a - b)^2$, and is therefore always a positive quantity. Thus, V_1 is not equal to V, but is always less than V.

How great may the discrepancy between V and V_1 become? The expression for the difference $V - V_1$ shows that it increases as the difference between a and b increases. The most unfavourable case will be the one when $b = 0$, i.e. when the upper base of the truncated pyramid becomes a single point and the pyramid from a truncated one turns into a complete one. In that case

$V = \frac{1}{3}a^2h$, $V_1 = \frac{1}{4}a^2h$, $V - V_1 = \frac{1}{12}a^2h = \frac{1}{4}V$, and the calculation, according to the formula for V_1, yields a value of the volume less than the actual one by 25 per cent of the latter.

In order to estimate the error in the general case, when $b \neq 0$, write $\dfrac{a-b}{a} = x$ and express the ratio $\dfrac{V - V_1}{V}$ as function of x. Since $b = a(1 - x)$, substitution yields, after simplification by a^2 and h, the formula $\dfrac{V - V_1}{V} = \dfrac{x^2}{12(1 - x + \frac{1}{3}x^2)}$. If x is a small fraction, its value may be neglected in relation to 1, and, *a fortiori*, we may neglect also the fraction $\frac{1}{3}x^2$. The whole expression within the brackets in the denominator of the fraction then becomes unity and we arrive at the formula:

$$\frac{V - V_1}{V} \approx \frac{1}{12}x^2, \quad \text{where} \quad x = \frac{a-b}{a}.$$

This formula asserts that when the difference between a and b is, for example, 0·1 of a, V_1 is less than V by approximately $\frac{1}{12}\ 0{\cdot}1^2 = \frac{1}{1200}$ of V. Indeed, taking $a = 10$ cm, $b = 9$ cm, $h = 12$ cm, we obtain according to the exact formula, $V = 1084$ cm^3, and by the approximate formula, $V_1 = 1083$ cm^3. The difference $V - V_1$ is equal to one cubic centimetre which constitutes about $\frac{1}{1100}$ of V.

For small values of the difference $a - b$ with respect to a, the approximate formula yields, as we see, fairly correct results.

Of course, the same method of approximate computation of the volume is applicable also to the computation of the volume of a truncated cone—take the product of the mean section of the truncated cone by the altitude. The mean section of the truncated cone is a circle with a radius (diameter) equal to half the sum of radii (diameters) of both bases of the truncated cone.

CHAPTER V

Trigonometry

I. Examples of False Arguments

69. $\frac{\pi}{4} = 0$

We proceed from an equality whose correctness is entirely obvious:

$$1 + \sin\frac{\pi}{4} - \cos\frac{\pi}{4} = \tan\frac{\pi}{4}, \qquad (1)$$

since

$$\sin\frac{\pi}{4} = \cos\frac{\pi}{4}, \quad \tan\frac{\pi}{4} = 1.$$

Applying the relation $\tan\frac{\pi}{4} = \frac{\sin\frac{\pi}{4}}{\cos\frac{\pi}{4}}$ and eliminating the denominator, we have:

$$\cos\frac{\pi}{4} + \sin\frac{\pi}{4}\cos\frac{\pi}{4} - \cos^2\frac{\pi}{4} = \sin\frac{\pi}{4},$$

whence

$$\cos\frac{\pi}{4}\left(\sin\frac{\pi}{4} - \cos\frac{\pi}{4}\right) = \sin\frac{\pi}{4} - \cos\frac{\pi}{4}. \qquad (2)$$

From relation (2) we establish that

$$\cos\frac{\pi}{4} = 1. \qquad (3)$$

Consequently, $$\frac{\pi}{4} = 2k\pi, \tag{4}$$

which, for $k = 0$, yields the required result:

$$\frac{\pi}{4} = 0. \tag{5}$$

70. The sine of an angle decreases if one adds 360° to that angle

Say α is an arbitrary angle between 0 and 180°. Its half, which we shall denote by the letter x, is consequently included between 0 and 90°. The sine and the cosine of that angle of the first quadrant are positive. Adding 180° to x we shall obtain an angle in the third quadrant, for which both the sine and the cosine, as is well known, are negative. But every negative quantity is less than a positive quantity, and therefore:

$$\sin (180° + x) < \sin x, \quad \cos (180° + x) < \cos x. \tag{1}$$

Multiplying these two inequalities term by term, we obtain the inequality:

$$\sin (180° + x) \times \cos (180° + x) < \sin x \times \cos x, \tag{2}$$

which may be rewritten in a shorter form, by making use of the formula for the sine of a double angle ($\sin 2\alpha = 2 \sin \alpha \times \cos \alpha$), namely:

$$\tfrac{1}{2} \sin (360° + 2x) < \tfrac{1}{2} \sin 2x.$$

Multiplying both members of the last inequality by 2 and replacing x by $\frac{1}{2}\alpha$, we finally have:

$$\sin (360° + \alpha) < \sin \alpha,$$

which "proves" the assertion enunciated in the title of the present problem and which contradicts the well-known fact that no trigonometric function changes with the increase of an angle by 360°.

71. The cosine of any acute angle is greater than unity

Taking an arbitrary acute angle α, write down the identity $\cos \alpha = \cos \alpha$ and take its logarithm (say, to the base ten). Thus:

$$\log \cos \alpha = \log \cos \alpha. \quad (1)$$

Replace the equality (1) by an inequality, doubling its left-hand member:

$$2 \log \cos \alpha > \log \cos \alpha, \quad (2)$$

or, what amounts to the same thing:

$$\log \cos^2 \alpha > \log \cos \alpha. \quad (3)$$

Taking into account, that when the base exceeds one, the greater the number the greater the logarithm, and conversely, from the inequality (3) we deduce that $\cos^2 \alpha > \cos \alpha$.

Dividing both members of the last inequality by the positive number $\cos \alpha$, we obtain an inequality in the same sense:

$$\cos \alpha > 1,$$

and arrive at a contradiction of the definition of the cosine of an acute angle, as the ratio of the adjacent side to the hypotenuse.

72. $\frac{1}{4} > \frac{1}{2}$

We proceed from the indubitable equality

$$\sin \frac{\pi}{6} = \sin \frac{\pi}{6}. \quad (1)$$

Taking the logarithms of both of its members, we write down

$$\log \sin \frac{\pi}{6} = \log \sin \frac{\pi}{6}. \quad (2)$$

From the relation (2) we pass to the inequality:

$$2 \log \sin \frac{\pi}{6} > \log \sin \frac{\pi}{6}, \quad (3)$$

which we rewrite in the form:

$$\log \sin^2 \frac{\pi}{6} > \log \sin \frac{\pi}{6}.$$

taking exponentials, we have:

$$\sin^2 \frac{\pi}{6} > \sin \frac{\pi}{6}. \qquad (4)$$

The inequality (4) asserts, that

$$(\tfrac{1}{2})^2 > \tfrac{1}{2}, \quad \text{i.e.} \quad \tfrac{1}{4} > \tfrac{1}{2}. \qquad (5)$$

73. $2^2 = 4^2$

Both members of the relation

$$\cos^2 x = 1 - \sin^2 x \qquad (1)$$

known from the trigonometry course, we raise to the power $\frac{3}{2}$:

$$\cos^3 x = (1 - \sin^2 x)^{3/2}. \qquad (2)$$

Increase each of the members of equality (2) by three unities and square the sums obtained:

$$(\cos^3 x + 3)^2 = [(1 - \sin^2 x)^{3/2} + 3]^2. \qquad (3)$$

Into relation (3) substitute a value of x, equal, for example, to $\frac{\pi}{2}$, and we obtain the true equality: $(0 + 3)^2 = [(1 - 1)^{3/2} + 3]^2$, i.e.

$$3^2 = 3^2.$$

However, if in that same relation we substitute a value of x, equal, for example, to π, we obtain the untrue equality:

$$(-1 + 3)^2 = [(1 - 0)^{3/2} + 3]^2,$$

i.e.

$$2^2 = 4^2.$$

Explain the reason for this apparent paradox.

74. The area of a rectangle is equal to zero

Take a trigonometric circle and on its circumference mark a point M representing the end of some arc AM of the first quadrant (Fig. 43). Drawing $MM_1 \perp AA_1$, $MQ \perp BB_1$, $M_1Q_1 \perp BB_1$, where AA_1 and BB_1 are a pair of mutually perpendicular diameters

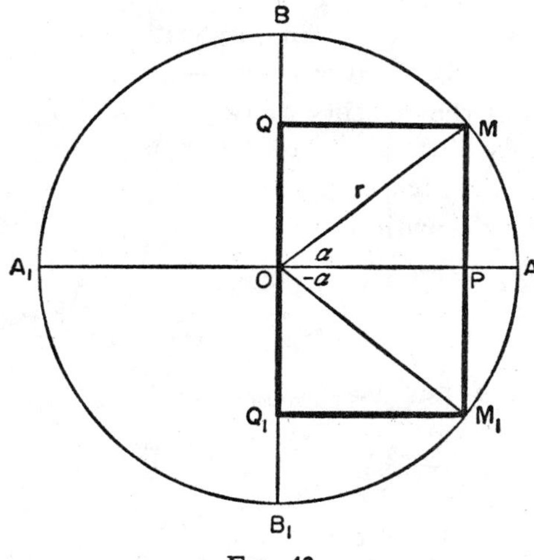

FIG. 43

of the trigonometric circle, we obtain the rectangle MQQ_1M_1 and proceed to calculate its area S. The base of the rectangle Q_1M_1 is equal to the segment OP, which is the "cosine line" for the angle α, and therefore is equal to $r \cos \alpha$, where r is the radius of the trigonometric circle. The altitude M_1M of the rectangle represents the sum of two segments PM and PM_1. The former segment is the sine line for the angle α and is equal to $r \sin \alpha$, while the second is the sine line for the angle $-\alpha$, and is equal to $r \sin(-\alpha) = -r \sin \alpha$. Hence we obtain:

$$M_1M = PM + PM_1, \tag{1}$$

$$M_1M = r \sin \alpha + (-r \sin \alpha) = r(\sin \alpha - \sin \alpha) = 0,$$

$$S = OP \times M_1M = r \cos \alpha \times 0 = 0.$$

75. There exist congruent triangles not all of whose sides are respectively equal

In order to convince oneself of congruence of two triangles, there is no need to know all the elements of these triangles. It is sufficient to be sure of the equality of some of them. In fact the first test for the congruence of triangles requires the equality of two sides and an angle, the second indication—of two angles and a side and, finally, the third indication—of three sides.

Thus, to assert the equality of two triangles it is required to know the equality of three of their elements, among which at least one linear element is represented.

Consider triangles with the sides:

$$a_1 = 18 \text{ cm}, \quad b_1 = 12 \text{ cm}, \quad c_1 = 8 \text{ cm},$$
$$a_2 = 27 \text{ cm}, \quad b_2 = 18 \text{ cm}, \quad c_2 = 12 \text{ cm}.$$

It is easy to see that the triangles of this pair have two pairs of equal sides.

By means of the formulae

$$\tan\frac{\alpha}{2} = \sqrt{\left(\frac{(p-b)(p-c)}{p(p-a)}\right)}, \qquad \tan\frac{\beta}{2} = \sqrt{\left(\frac{(p-a)(p-c)}{p(p-b)}\right)},$$
$$\tan\frac{\gamma}{2} = \sqrt{\left(\frac{(p-a)(p-b)}{p(p-c)}\right)}$$

we establish, that for each of the angles of one triangle there will be one equal to it among the angles of the other.

In fact:

$$p_1 = \frac{a_1 + b_1 + c_1}{2} = 19, \quad p_1 - a_1 = 1, \quad p_1 - b_1 = 7,$$
$$p_1 - c_1 = 11; \quad p_2 = \frac{a_2 + b_2 + c_2}{2} = 28{\cdot}5,$$
$$p_2 - a_2 = 1{\cdot}5, \quad p_2 - b_2 = 10{\cdot}5, \quad p_2 - c_2 = 16{\cdot}5.$$

Consequently,

$$\tan\frac{\alpha_1}{2} = \sqrt{\left(\frac{7\times 11}{19}\right)} = \sqrt{\left(\frac{10\cdot 5\times 16\cdot 5}{1\cdot 5\times 28\cdot 5}\right)} = \tan\frac{\alpha_2}{2};$$

$$\tan\frac{\beta_1}{2} = \sqrt{\left(\frac{11}{7\times 19}\right)} = \sqrt{\left(\frac{16\cdot 5\times 1\cdot 5}{10\cdot 5\times 28\cdot 5}\right)} = \tan\frac{\beta_2}{2};$$

$$\tan\frac{\gamma_1}{2} = \sqrt{\left(\frac{7}{11\times 19}\right)} = \sqrt{\left(\frac{1\cdot 5\times 10\cdot 5}{16\cdot 5\times 28\cdot 5}\right)} = \tan\frac{\gamma_2}{2}.$$

In other words,

$$\alpha_1 = \alpha_2, \quad \beta_1 = \beta_2, \quad \gamma_1 = \gamma_2.$$

Thus, five elements of one triangle, among them two linear, are equal to five elements of the other triangle. Hence we conclude as to the equality of these triangles.

Thus, there exist congruent triangles in which not all the sides are equal.

76. Every triangle is a right-angled triangle

ABC is an arbitrarily taken triangle with sides a, b, c, angles α, β, γ and altitude h, which is dropped to the side c and divides it into segments p and q.

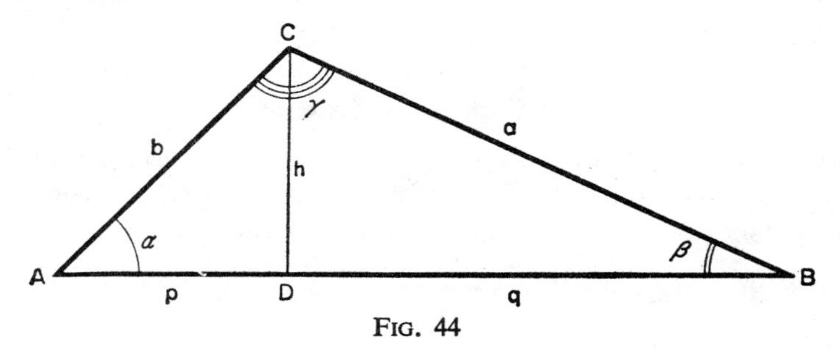

FIG. 44

From the diagram we establish without difficulty:

$$\sin\alpha = \frac{h}{b}, \quad \cos\alpha = \frac{p}{b}, \quad \sin\beta = \frac{h}{a} \text{ and } \cos\beta = \frac{q}{a}. \qquad (1)$$

Making use of the relations (1) we may give to the formula:

$$\sin(\alpha + \beta) = \sin \alpha \cos \beta + \cos \alpha \sin \beta \tag{2}$$

a somewhat different form, namely:

$$\sin(\alpha + \beta) = \frac{h}{b} \times \frac{q}{a} + \frac{p}{b} \times \frac{h}{a} = \frac{h(p + q)}{ab} = \frac{hc}{ab}. \tag{3}$$

For further transformations of formula (3) we recall that:

$$a = 2R \sin \alpha, \quad b = 2R \sin \beta, \quad c = 2R \sin \gamma,$$

where R is the radius of the circle circumscribed about the triangle. Substitution of these relations in the right-hand member of formula (3) transforms it into the form:

$$\sin(\alpha + \beta) = \frac{h \sin \gamma}{R \sin \alpha \sin \beta}. \tag{4}$$

Finally, noting, that $h = b \sin \alpha = 2R \sin \alpha \sin \beta$, we arrive at the conclusion:

$$\sin(\alpha + \beta) = \sin \gamma, \tag{5}$$

whence

$$\alpha + \beta = \gamma, \tag{6}$$

which, obviously, can hold only when γ is equal to a right angle.

II. Analysis of the Examples

69. The error is introduced in passing from the true equality (2) to the incorrect equality (3). The falsity of equality (3) is explained by the fact that it is obtained as result of dividing both members of equality (2) by the difference $\sin \frac{\pi}{4} - \cos \frac{\pi}{4}$, which is equal to zero.

70. As a result of a term-by-term multiplication of the two inequalities in the same sense, all the terms of which are positive, we obtain a new inequality in the same sense. However, multiplication of inequalities not satisfying this requirement may yield an arbitrary result: either an inequality in the other sense, or even an equality.

In the inequalities (1) the left- and the right-hand members are equal in absolute value, but differ in sign:

$$\sin (180° + x) = -\sin x \text{ and } \cos (180° + x) = -\cos x.$$

Their term-by-term multiplication leads to two products equal in absolute value and having the sign plus, and therefore equal to each other. The inequality (2) is therefore incorrect and should be replaced by the equality:

$$\sin (180° + x) \times \cos (180° + x) = \sin x \times \cos x,$$

which leads to the well-known formula:

$$\sin (360° + 2x) = \sin 2x \text{ and } \sin (360° + \alpha) = \sin \alpha.$$

71. In passing from equality (1) to inequality (2) we have multiplied by 2 the left-hand member of equality (1).

But is this multiplication by 2 really an increase? Is always $2a > a$? Transferring in the latter equality the number a from the right-hand member to the left, we convince ourselves that from the inequality $2a > a$ follows the inequality $a > 0$. Consequently, if a is not greater than zero but less than, or equal, to zero, then the inequality $2a > a$ cannot hold.

Thus, the passage from the equality $a = a$ to the inequality $2a > a$ is possible exclusively for $a > 0$. But the cosine of an acute angle is always included between zero and unity, and therefore from the equality $\log \cos \alpha = \log \cos \alpha$, does not follow the inequality

$$2 \log \cos \alpha > \log \cos \alpha.$$

72. The mistake is introduced on passing from equality (2) to inequality (3). Here it has escaped notice that, since $\sin \frac{\pi}{6} = \frac{1}{2}$, i.e. a proper fraction, $\log \sin \frac{\pi}{6}$ is negative and thus in relation (3) we should put not the symbol greater than but, inversely, the symbol less than: the double of a negative number is less than that negative number itself.

73. The error is based on forgetting the definition of an arithmetic root. In correspondence with that definition $\sqrt{(a^2)} \neq a$, but is equal to $|a|$, i.e. $\sqrt{(a^2)} = a$ if $a \geqslant 0$, and $\sqrt{(a^2)} = -a$, if $a < 0$.

In the example under analysis: $(\cos^2 x)^{3/2} = \sqrt{(\cos^6 x)} = \cos^3 x$, if

$$-\frac{\pi}{2} + 2k\pi \leqslant x \leqslant \frac{\pi}{2} + 2k\pi, \text{ and } \sqrt{(\cos^6 x)} = -\cos^3 x,$$

if x does not belong to the set of values indicated by the double inequality. In particular, when x is equal to π, the left-hand member of the relation (3) should be taken in the form:

$$(-\cos^3 x + 3)^2.$$

The absurd conclusion is then eliminated.

74. The error in the present argument consists, of course, in the incorrect calculation of the altitude of the rectangle. Ascribing to the segment a definite sign (+ or −), we indicate in which direction this segment is laid off, or, what amounts to the same thing, in which direction we traverse that segment. For the "line of sine," the sign plus indicates that it is laid off upward from the horizontal diameter, and minus—downward. Considering such a direction of the segment and denoting each segment by two letters, we always assume that the first letter indicates the beginning of the segment, and the second its end. Therefore the two directed segments AB and BA are not identical: having the same length, they have opposite directions, and, by virtue of this, $AB = -BA$.

On Fig. 43 the segment PM is laid off upward, and therefore $PM = + \sin \alpha$, while the segment PM_1 is laid off downward, and consequently, $PM_1 = -r \sin \alpha$. In order to obtain the altitude of the rectangle with the sign plus, we have to move from the point M_1 to the point M in one and the same direction, namely upward. Therefore the formula (1) is incorrect, while the correct one is formula $M_1M = M_1P + PM$ or, what amounts to the same, $M_1M = PM + M_1P$. Noting that $M_1P = -PM_1 = -(-r \sin \alpha) = +r \sin \alpha$, we have:

$$M_1M = r \sin \alpha + r \sin \alpha = 2r \sin \alpha,$$

as should be the case.

The present sophism could have been explained in a much shorter way, simply by indicating that in calculating the area of a rectangle we should take only the lengths of its base and altitude without taking their signs into account. However, definite signs (+ and −) are attached occasionally even to areas; also, it is important to know how, by directed segments from some given point to the ends of some other segment, to obtain the length of this segment. (Our problem consists in the following: we had to find the length of segment M_1M, knowing the directed segments leading from point P to points M_1 and M.)

75. On the basis of the analysis of the three tests for congruence of triangles by the method of incomplete induction, a false conclusion is arrived at: "Thus, to assert the congruence of two triangles, it is necessary to know (i.e. sufficient to know!) that there exists the equality of three of their elements, among which at least one linear element is represented."

In fact, however, the equality of three fundamental elements of one triangle to three fundamental elements of another triangle, including at least one linear element, is a necessary, but not a sufficient, condition for congruence of the figures under consideration. By examining Fig. 45, we easily convince ourselves of the insufficiency of that condition.

Taking the false assertion for a true one and applying it in the analysis of the example, in which we are dealing with similar triangles with sides 18, 12, 8 and 27, 18, 12, we came to the assertion of an absurdity—there exist congruent triangles in which not all sides are equal.

This sophism was considered earlier in geometry (Chapter IV). There its detailed solution was given.

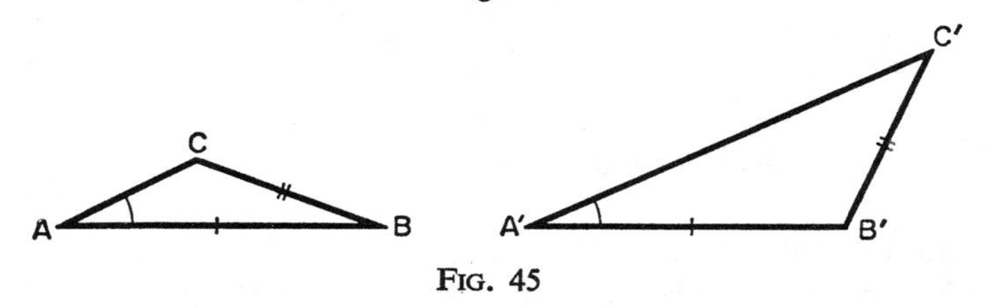

FIG. 45

76. The mistake is committed in passing from relation (5) to relation (6). It consists in the incorrect application of the principle of direct conclusions by the converse. The fact that equal angles have equal sines still does not allow us to conclude the correctness of the converse assertion. Here we may assert only the following: "if the sines of two angles are equal, then also the angles may be equal."

With respect to the angles $(\alpha + \beta)$ and γ three assumptions may be made:

(1) $\alpha + \beta = \gamma$. In this special case we shall actually have a right triangle. But this case, as may be seen from what follows, does not exhaust the realm of all possible cases.

(2) $a + \beta = 180° - \gamma$. In that case $\alpha + \beta + \gamma = 180°$, which does not contradict the sense of the problem and relates the angles of the triangle by the known relationship. Moreover, this case gives no basis for the absurd conclusion of the presence of a right angle in every triangle.

(3) $\alpha + \beta = 360° + \gamma$. This case is impossible, since the difference of the sum of two arbitrary angles of a triangle and its third angle cannot be equal to 360°.

As we see, the absurd conclusion is eliminated.

CHAPTER VI

Approximate Computations

Accounts Explaining Sources of Erroneous Reasoning

77. How old is the ancient statue?

To the question, how old is the ancient statue in the museum, the museum attendant answered: "It is 4008 years old." Thereupon followed the question: "How was the age of such an ancient statue determined with such high degree of accuracy?" The attendant explained that he began to work in the museum eight years ago and then learned from the museum director that the statue was 4000 years old. Eight years had passed since then, and therefore the age of the statue at the present time was 4000 + 8 = 4008 (years).

What should one think of such an explanation?

The museum attendant obviously did not understand that the age of the statue is determined as four thousand years not exactly, but only approximately. It is known how many milleniums have passed since the time the statue was made, but how many centuries have passed since then, how many decades, not to speak of single years, remains unknown. Replacing the unknown figures in the age indicated by the director by question marks, we should write the addition carried out by the attendant in the following form:

$$\begin{array}{r} 4??? \\ +\quad 8 \\ \hline 4008 \end{array}$$

But it is quite clear, that from adding 8 to the number expressed by an unknown number of units of the first component, one may obtain a number with an arbitrary number of units, and not at all necessarily with the number 8. In exactly the same way, nothing justifies the putting of zeros as the tens and hundreds of the sum. The only correct result will be the following:

$$\begin{array}{r} 4??? \\ +\quad 8 \\ \hline 4??? \end{array}$$

Consequently, the age of the statue, initially determined as four thousand years, remains the same after a lapse of eight years.

In adding and subtracting numbers known not with absolute accuracy but only to some approximation, we should always clarify which digits of the result should be given credence, and which not, the latter to be altogether discarded.

Suppose a box with an instrument has been received. On the box there is the legend: gross weight (i.e. weight of the merchandise together with the packing) 35 kg. Say, the weight of the instrument itself (the net weight) turns out, upon weighing on an exact scale, to be equal to 2·853 kg. The conclusion that the packing weighs $35 - 2{\cdot}853 = 32{\cdot}147$ (kg), is incorrect. All the numbers after the decimal point deserve no credence and the result should be rounded off to the integers: the packing weighs 32 kg.

78. All large numbers are approximately equal to each other

We shall agree to consider numbers from one million onwards as large, and shall demonstrate that, for example, $1{,}000{,}000 \approx 2{,}000{,}000$.

The assertion that for every large number N one may consider

$$N \approx N + 1, \tag{1}$$

calls forth no protests.

Successively substituting in the relation (1) 1,000,000, 1,000,001, 1,000,002, . . ., 1,999,999, we have:

$$
\begin{aligned}
&1{,}000{,}000 \approx 1{,}000{,}001; \\
&1{,}000{,}001 \approx 1{,}000{,}002; \\
&1{,}000{,}002 \approx 1{,}000{,}003; \\
&\ldots\ldots\ldots\ldots \\
&\ldots\ldots\ldots\ldots \\
&\ldots\ldots\ldots\ldots \\
&1{,}999{,}999 \approx 2{,}000{,}000. \qquad (2)
\end{aligned}
$$

Multiplying together the left and the right-hand members of the million approximate equalities (2), we obtain

$$
\begin{aligned}
&1{,}000{,}000 \times 1{,}000{,}001 \times 1{,}000{,}002 \times \ldots \times 1{,}999{,}999 \\
&\quad \approx 1{,}000{,}001 \times 1{,}000{,}002 \times 1{,}000{,}003 \times \ldots \\
&\qquad \times 1{,}999{,}999 \times 2{,}000{,}000 \qquad (3)
\end{aligned}
$$

Simplifying both members of the approximate equality (3) by

$$1{,}000{,}001 \times 1{,}000{,}002 \times 1{,}000{,}003 \times \ldots \times 1{,}999{,}999,$$

we obtain the required relation:

$$1{,}000{,}000 \approx 2{,}000{,}000.$$

What is the matter?

The term-by-term multiplication of equalities belongs to the set of lawful mathematical operations. However, it is exact equalities that are meant. As far as approximate equalities are concerned, they represent, in their mathematical sense, inequalities: the notation $x \approx a$, with an accuracy of 10^{-n}, is equivalent to the notation $a - 10^{-n} < x < a + 10^{-n}$, and with an accuracy of 10^{+n} to the notation $a - 10^{n} < x < a + 10^{n}$.

It is clear from the above that the term-by-term multiplication of approximate equalities should be treated as the multiplication of the corresponding inequalities. But, as is well known, in a term-by-term multiplication of inequalities in the same sense (and only such may be multiplied together) the inequality is strengthened. In fact, having the inequalities $a - \alpha < x < a + \alpha$ and $b - \beta < y < b + \beta$, we conclude that $ab - \alpha b - \beta a + \alpha\beta < xy < ab + \alpha b + \beta a + \alpha\beta$, i.e. the multiplication of two double inequalities strengthens each of them.

It is natural that in multiplying a large number of inequalities (in the case under analysis—one million) the inequality is greatly strengthened.

Thus, the absurd conclusion is eliminated.

This sophism is constructed on the basis of the ambiguity of the term "equality": exact equality and approximate equality.

79. On the precision of the product of approximate numbers

Say one is to find the product of two numbers $124\frac{1}{3}$ and $27\frac{10}{73}$. Not wishing to deal with ordinary fractions, we pass to decimal ones, limiting ourselves in accuracy to tenths. Since $\frac{1}{3} = 0{\cdot}33\ldots$, $\frac{10}{73} = 0{\cdot}13\ldots$, we carry out the multiplication of the decimal numbers 124·3 and 2·1, and obtain the product 261·03. The multipliers were taken with an accuracy up to tenths. Considering that the result is obtained also with the same accuracy, we finally write:

$$124\tfrac{1}{3} \times 2\tfrac{10}{73} = 261{\cdot}0.$$

Is this correct?

Carrying out a multiplication without the preliminary transformation of the factors into decimal fractions, we find that the product is equal to $265\frac{51}{73}$, or 265·698.... Consequently, in spite of our expectations, there is an error not only in the decimal fraction digit, but also in the number of units.

In order to understand how this arose, we write the approximate multipliers in the form 124·3? and 2·1?, replacing by question marks the unknown digits of the one-hundredths fractions, and we carry out the multiplication just as if these hundredths digits were known. Multiplying the unknown digits in the partial products, at the corresponding places we also set question marks:

$$\begin{array}{rr}
 & 124{\cdot}3\,? \\
\times & 2{\cdot}1\,? \\
\hline
 & ????? \\
1 \;\vdots & 243\,? \\
24 \;\vdots & 86\,? \\
\hline
26 \;\vdots & 1{\cdot}03\,??
\end{array}$$

All the digits of the product to the right of the vertical dotted line are obtained from the addition of unknown numbers to each other, and therefore deserve no credence. Discarding these digits, we write down the product in the form 260, or, even better, in the form 26_0: the reduced size of the zero in the units digit indicating that this zero is put instead of a units digit which remains unknown.

When multiplying approximate numbers one has to be guided by the rule: in a product of two approximate numbers, retain a number of significant figures equal to the number of significant figures in the factor containing the smaller number of them. By significant figures we mean all of its digits, i.e. both the digits of the integral part and those of the fractional part, with the exception of zeros which stand to the left of the first non-zero digit, and of those zeros which may stand to the right of the number, if those zeros are put instead of unknown digits (for example, the numbers 20·6, 206, 0·00206 each have three significant figures).

Applied to the example (analysed above) of the multiplication of approximate numbers, this rule recommends the retention in the product of only the first two significant numbers, since the approximate multiplicand (124·3) has four significant figures, but the approximate multiplier (2·1) has only two. But we have convinced ourselves that only these two first digits of the product are correct in it. Taking the multiplier $2\frac{10}{73} = 2{\cdot}1370\ldots$ in the form of a decimal fraction not with two but with three significant figures (2·14) we should have obtained a product 266·002, in which (according to the rule) one would have to retain the first three significant figures; and we should have obtained the result $124\frac{1}{3} \times 2\frac{10}{73} \approx 266$, i.e. precisely what is obtained if the exact product 265·698 is rounded off to three significant figures. If we took the multiplier with four significant figures (2·137) we should obtain the product 265·6291, in which, according to the rule, we should have to retain the first 4 significant figures. And, indeed, the discrepancy between this approximate product and its exact value (265·698 . . .) begins only after the fourth significant figure.

We make use of this very convenient rule for rounding off the product of approximate numbers also when one of the factors is an exact number. Then we merely have to consider that in that factor there is an infinitely great number of significant figures and retain in the product only as many significant figures as there are in the second (approximate) factor. For example, if we take $\pi \approx 3{\cdot}14$ and find the length of a circumference corresponding to a diameter equal to 2·6 cm, then in the product $3{\cdot}14 \times 2{\cdot}6 = 8{\cdot}164$, we should retain the first 3 significant figures, which are the only true digits of this product. (If we took $\pi = 3{\cdot}14159$, we should obtain $\pi \times 2{\cdot}6 = 8{\cdot}1681 \ldots$.)

It should be noted that there may occur cases when the product of two numbers has more correct digits than may be expected on the basis of this rule, since one approximate factor may be less than the corresponding exact value, while the other may be greater. It may also just happen that the last digit, which we retain in the product according to the rule, will not be totally exact. Theory indicates that in this last digit there may be an imprecision, to the extent, in exceptional cases, of 6 units in that digit, but never exceeding that limiting value.*

Similar rules should be observed also in carrying out other

* About details relating to the derivation and application of this rounding off rule for approximate numbers, see the book by V. M. Bradis, *Means and Methods for Elementary Calculations* (Sredstva i sposoby elementarnykh ischislenii, § 13, Moscow, 1954.

operations on approximate numbers—in the quotient of two approximate numbers we should retain as many significant figures as there are in that one of the two numbers given (dividend and divisor) which has the smaller number of significant figures. When raising a number to the square, or cube, we should retain as many significant figures in the result as there are in the approximate number being raised to a power. The square and cube roots of approximate numbers should be extracted with the same number of significant figures as there are in the numbers under the radical sign.*

80. Is the formula $\frac{\sin \alpha}{\cos \alpha} = \tan \alpha$ true?

A student, having come across the formula $\frac{\sin \alpha}{\cos \alpha} = \tan \alpha$ decided to check it. Copying out from the table the values of all three functions for $\alpha = 89° 42'$, namely: $\sin \alpha = 1{\cdot}0000$, $\cos \alpha = 0{\cdot}0052$, $\tan \alpha = 191{\cdot}0$, he divided $\sin \alpha = 1{\cdot}0000$ by $\cos \alpha = 0{\cdot}0052$, but obtained not the number 191·0, as he expected, but the number 192·3..., and he concluded that either the formula was untrue or at least one of the three tabulated values was incorrect.

Of course, the reason for the discrepancy is the fact that all the tabulated values written out, just as the overwhelming majority of other tabulated values, are *approximate* numbers, giving only the first few significant figures corresponding to the exact values. In dividing two approximate numbers the quotient is obtained not exactly, but only approximately. The rules dealt with in Problem 79 indicate how many significant figures should be retained in such a quotient. In the present case the dividend (1·0000) has five significant figures, the divisor (0·0052) has only two, and therefore also in the quotient we should retain only the first two significant figures. Having obtained the quotient

* For details relating to this, see Bradis, *loc. cit.*, § 14 and 19.

192·3... we round it off so that there remain only the first two significant figures, and obtain the number 190, which coincides as to the first two significant figures with the tabulated value of tan 89° 42′, equal to 191·0.

In order to obtain a quotient with four significant figures one has to increase the precision of the divisor, taking it not with two, as in our case, but with at least four significant figures (in order to obtain four completely reliable significant figures in the quotient, the dividend and the divisor should rather be taken with one "extra" figure, i.e. not with four, but with five significant figures). Thus, finding (according to a more precise table), that cos 89° 42′ = 0·0052360, we divide 1·0000 by 0·0052360, and obtain in the quotient 190·98... or, upon rounding off to four significant figures, precisely the tabulated value of tan 89° 42′, namely 191·0.

81. How many digits should be known in a number under a radical sign, in order to obtain the root with a given accuracy?

Assume, that we wish to compute the value of sin 15° according to the formula $\sin 15° = 0{\cdot}5\sqrt{(2 - \sqrt{3})}$, this value to be correct to four significant figures. With what accuracy is it necessary to extract $\sqrt{3}$, in order to ensure the required accuracy of the result in the second extraction of the square root of $(2 - \sqrt{3})$? It is often argued thus—in order to obtain each next figure of the square root, we require a new digit for the pair of figures in the pair of numbers under the radical sign. Consequently, in order to obtain four figures after the decimal sign in the value of the root of $2 - \sqrt{3}$, it is necessary in this latter number to have $2 \times 4 = 8$ figures after the decimal sign. Guiding ourselves by this consideration, the square root of 3 is extracted with eight decimals, and, obtaining $\sqrt{3} = 1{\cdot}73205081$, we find $2 - \sqrt{3} = 0{\cdot}26794919$. Thereupon we compute $\sqrt{(2 - \sqrt{3})} = 0{\cdot}5176$ and, finally, $\sin 15° = 0{\cdot}5\sqrt{(2 - \sqrt{3})} = 0{\cdot}2588$.

As shown by a check in a table of sines, all the four decimal digits of this result are correct. But the same result is also obtained by the following calculation, requiring a considerably smaller number of operations:

$$\sqrt{3} = 1{\cdot}7321,$$
$$2 - \sqrt{3} = 0{\cdot}2679,$$
$$\sqrt{(2 - \sqrt{3})} = 0{\cdot}5176,$$
$$\sin 15^\circ = 0{\cdot}5\sqrt{(2 - \sqrt{3})} = 0{\cdot}2588.$$

This example leads to the idea that, in order to obtain each extra number in the value of a square root, there is no need to know the two extra digits in the number under the radical sign. Indeed, it is not difficult to show the expediency of such a rule— in extracting the square root of an approximate number having n significant figures, one should also take n significant figures in the root.* The last digit of the root so obtained will then be either totally exact, or differ from the exact value by one unit. Therefore, in order to obtain the square root to n significant figures, it is sufficient to take the number under the radical sign also with n significant figures (or, if it is desired to have a complete guarantee of accuracy of the last figure, then with $n + 1$ significant figures).

Consider another example. Say it is required to find the value of $x = \sqrt{(28\frac{2}{7})}$ with an accuracy up to the hundredths. Noting that the root contains only one number in the integral part, we see that its value has to be found with three significant figures (the number of whole units, the decimal digit and the hundredths digit). Therefore, it is also sufficient to take the number $28\frac{2}{7}$ with three significant figures, i.e. in the form 28·3. Extracting the root of that latter number, we have the final result 5·32. For checking, we find the same root by another method, namely reducing the problem to the extraction of a root of an integer:

$$\sqrt{(28\tfrac{2}{7})} = \sqrt{(\tfrac{198}{7})} = \tfrac{1}{7}\sqrt{(1386)} = \tfrac{1}{7} \times 37{\cdot}229\ldots = 5{\cdot}318\ldots \approx 5{\cdot}32.$$

* See Bradis, *loc. cit.*, § 19.

82. Why does one get rid of irrationals in the denominator?

Often the necessity for eliminating irrationals in the denominator of a fractional expression is explained by the fact that, due to this, an increase in accuracy is attained. Thus, in the reputable book by P. I. Shmulevich "Dopolneniya k kursu algebry trebuemye programmamy konkursnykh ekzamenov" (Supplements to the Algebra Course, Required by Competitive-Exam. Curriculum) 10th ed., 1917, we find (on p. 63) the assertion, that in calculating the $\frac{1}{n}$ th root, we shall obtain the numerical value of the expression $\frac{a}{\sqrt[m]{b}}$ with b-fold precision, if we first transform it into the form

$$\frac{a\sqrt[m]{(b^{m-1})}}{b}.$$

Is it really so?

Let us consider an example: say $x = \frac{a}{\sqrt[m]{b}}$, where $a = 1$, $b = 10$, $m = 2$. We are to find the value of x, extracting the root with an accuracy up to tenths. Substituting the numerical values directly, we have $x_1 = \frac{1}{\sqrt{10}} \approx \frac{1}{3{\cdot}2} = 0{\cdot}3125$. By performing the preliminary elimination of the irrational in the denominator, we obtain $x_2 = \frac{\sqrt{10}}{10} \approx \frac{3{\cdot}2}{10} = 0{\cdot}32$. Comparing both approximate values with more exact ones, equal to $0{\cdot}1\sqrt{10} = 0{\cdot}1 \times 3{\cdot}1622\ldots = 0{\cdot}31622\ldots$, we see that the first of them has an error of $0{\cdot}31622\ldots - 0{\cdot}3125 = 0{\cdot}00372\ldots$ by way of *deficiency* while the second (x_2) has an error of $0{\cdot}032 - 0{\cdot}31622\ldots = 0{\cdot}00378\ldots$ by way of *excess*. As we see, in spite of the assertion in the book, the elimination of the irrationality in the denominator in no way increased the accuracy by a factor of 10, but rather decreased it a little.

It is easy to show that in replacing the roots by their approximate values, differing from the exact values by less than $\frac{1}{n}$ of the unit digit, we obtain a value of the expression $\frac{a}{\sqrt[m]{(b)}}$, with an error not greater than $\frac{a}{\sqrt[m]{(b^2)}}$, and the value of the expression $\frac{a}{b}\sqrt[m]{(b^{m-1})}$ with an error not greater than $\frac{a}{nb}$. These bounds of the error for $m = 2$ are equal, which precisely confirms the example just considered. For $m > 2$ and $b > 1$, the second bound of the error is less than the first, but not b times, as asserted in the book, but only $\sqrt[m]{(b^{m-2})}$ times. For $m > 2$ and $b < 1$, the second bound is greater than the first.

Consider another example. Suppose that it is required to find the value of the expression $x = \frac{1}{\sqrt{a} + \sqrt{b}}$ for $a = 1{\cdot}02$, $b = 1{\cdot}01$, having at our disposal a four-figure table of square roots. The table yields $\sqrt{a} = 1{\cdot}010$, $\sqrt{b} = 1{\cdot}005$, and, carrying out the substitution directly in the given expression, we obtain

$$x_1 = \frac{1}{2{\cdot}015} = 0{\cdot}4963.$$

If we first transform the given expression, eliminating (by multiplying the numerator and denominator by $\sqrt{a} - \sqrt{b}$) the irrationality in the denominator, we obtain:

$$x_2 = \frac{\sqrt{a} - \sqrt{b}}{a - b} = \frac{\sqrt{1{\cdot}02} - \sqrt{0{\cdot}01}}{0{\cdot}01} \approx \frac{1{\cdot}010 - 1{\cdot}005}{0{\cdot}01} = 0{\cdot}5.$$

Whereas the first calculation gave a value of x with four decimal digits, the second gave it with only one, i.e. considerably less precisely.

As we see, the elimination of the irrationality in the denominator does not greatly increase the accuracy of the result of the calculation. The fact is, it even reduces it at times.

Why then do they carry on such a "struggle with irrationals in the denominator"?

The point is that the calculation is carried out in a great majority of cases much more conveniently, if there are no roots in the denominators. For example, for the calculation of $\frac{1}{\sqrt{10}}$ we have to carry out a division by a multidigit approximate value of the root, while after eliminating the irrationality in the denominator the division is carried out much simply: ($\sqrt{10} \div 10$).

Try to simplify the expression:

$$\frac{1}{\sqrt{(a+1)}+\sqrt{(a)}}+\frac{1}{2\sqrt{(a+2)}-\sqrt{(4a+2)}}+$$
$$+\frac{1}{\sqrt{(3a+1)}+\sqrt{(3a)}},$$

by first eliminating the roots in the denominator of every fraction, and then without resorting to that method; the benefit of the "struggle with irrationals in the denominator" will become quite evident. We need not think, however, that this struggle need always be carried on. In many problems in higher mathematics (in finding limits, in integrating irrational expressions, etc.) one is often compelled to do the opposite, i.e. to eliminate the irrationals in the numerator, transferring them to the denominator. Here is a simple example: to establish what occurs to the difference

$$\sqrt{(x+1)}-\sqrt{(x)}$$

as x increases without bound, i.e. to compute

$$\lim_{x\to\infty} \sqrt{(x+1)}-\sqrt{(x)}.$$

From the elements of the theory of limits, studied in the course of algebra at the ninth grade, it is known that the limit of the difference of two variables, having limits, is equal to the difference of their limits.

It is easy to see that in this example one may not apply the theorem on the limit of a difference, since no limits of the minuend and the subtrahend exist (the expression $\infty - \infty$ simply has no meaning).

However, we can easily solve the problem, if we make use of the following transformations of the expression found under the limit sign:

$$\lim_{x\to\infty} \frac{\sqrt{(x+1)} - \sqrt{x}}{1} = \lim_{x\to\infty} \frac{(\sqrt{(x+1)} - \sqrt{x})(\sqrt{(x+1)} + \sqrt{x})}{\sqrt{(x+1)} + \sqrt{x}}$$

$$= \lim_{x\to\infty} \frac{x+1-x}{\sqrt{(x+1)} + \sqrt{x}} = \lim_{x\to\infty} \frac{1}{\sqrt{(x+1)} + \sqrt{x}} = 0.$$

A CATALOG OF SELECTED

DOVER BOOKS

IN ALL FIELDS OF INTEREST

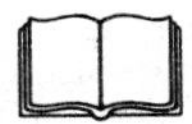

A CATALOG OF SELECTED DOVER BOOKS IN ALL FIELDS OF INTEREST

FRANK LLOYD WRIGHT'S HOLLYHOCK HOUSE, Donald Hoffmann. Lavishly illustrated, carefully documented study of one of Wright's most controversial residential designs. Over 120 photographs, floor plans, elevations, etc. Detailed perceptive text by noted Wright scholar. Index. 128pp. 9¼ x 10¾. 27133-1 Pa. $11.95

THE MALE AND FEMALE FIGURE IN MOTION: 60 Classic Photographic Sequences, Eadweard Muybridge. 60 true-action photographs of men and women walking, running, climbing, bending, turning, etc., reproduced from rare 19th-century masterpiece. vi + 121pp. 9 x 12. 24745-7 Pa. $10.95

1001 QUESTIONS ANSWERED ABOUT THE SEASHORE, N. J. Berrill and Jacquelyn Berrill. Queries answered about dolphins, sea snails, sponges, starfish, fishes, shore birds, many others. Covers appearance, breeding, growth, feeding, much more. 305pp. 5¼ x 8¼. 23366-9 Pa. $9.95

GUIDE TO OWL WATCHING IN NORTH AMERICA, Donald S. Heintzelman. Superb guide offers complete data and descriptions of 19 species: barn owl, screech owl, snowy owl, many more. Expert coverage of owl-watching equipment, conservation, migrations and invasions, etc. Guide to observing sites. 84 illustrations. xiii + 193pp. 5⅜ x 8½. 27344-X Pa. $8.95

MEDICINAL AND OTHER USES OF NORTH AMERICAN PLANTS: A Historical Survey with Special Reference to the Eastern Indian Tribes, Charlotte Erichsen-Brown. Chronological historical citations document 500 years of usage of plants, trees, shrubs native to eastern Canada, northeastern U.S. Also complete identifying information. 343 illustrations. 544pp. 6½ x 9¼. 25951-X Pa. $12.95

STORYBOOK MAZES, Dave Phillips. 23 stories and mazes on two-page spreads: Wizard of Oz, Treasure Island, Robin Hood, etc. Solutions. 64pp. 8¼ x 11. 23628-5 Pa. $2.95

NEGRO FOLK MUSIC, U.S.A., Harold Courlander. Noted folklorist's scholarly yet readable analysis of rich and varied musical tradition. Includes authentic versions of over 40 folk songs. Valuable bibliography and discography. xi + 324pp. 5⅜ x 8½. 27350-4 Pa. $9.95

MOVIE-STAR PORTRAITS OF THE FORTIES, John Kobal (ed.). 163 glamor, studio photos of 106 stars of the 1940s: Rita Hayworth, Ava Gardner, Marlon Brando, Clark Gable, many more. 176pp. 8⅜ x 11¼. 23546-7 Pa. $14.95

BENCHLEY LOST AND FOUND, Robert Benchley. Finest humor from early 30s, about pet peeves, child psychologists, post office and others. Mostly unavailable elsewhere. 73 illustrations by Peter Arno and others. 183pp. 5⅜ x 8½. 22410-4 Pa. $6.95

YEKL and THE IMPORTED BRIDEGROOM AND OTHER STORIES OF YIDDISH NEW YORK, Abraham Cahan. Film Hester Street based on Yekl (1896). Novel, other stories among first about Jewish immigrants on N.Y.'s East Side. 240pp. 5⅜ x 8½. 22427-9 Pa. $6.95

SELECTED POEMS, Walt Whitman. Generous sampling from *Leaves of Grass*. Twenty-four poems include "I Hear America Singing," "Song of the Open Road," "I Sing the Body Electric," "When Lilacs Last in the Dooryard Bloom'd," "O Captain! My Captain!"–all reprinted from an authoritative edition. Lists of titles and first lines. 128pp. 5$\frac{3}{16}$ x 8¼. 26878-0 Pa. $1.00

THE BEST TALES OF HOFFMANN, E. T. A. Hoffmann. 10 of Hoffmann's most important stories: "Nutcracker and the King of Mice," "The Golden Flowerpot," etc. 458pp. 5⅜ x 8½. 21793-0 Pa. $9.95

FROM FETISH TO GOD IN ANCIENT EGYPT, E. A. Wallis Budge. Rich detailed survey of Egyptian conception of "God" and gods, magic, cult of animals, Osiris, more. Also, superb English translations of hymns and legends. 240 illustrations. 545pp. 5⅜ x 8½. 25803-3 Pa. $13.95

FRENCH STORIES/CONTES FRANÇAIS: A Dual-Language Book, Wallace Fowlie. Ten stories by French masters, Voltaire to Camus: "Micromegas" by Voltaire; "The Atheist's Mass" by Balzac; "Minuet" by de Maupassant; "The Guest" by Camus, six more. Excellent English translations on facing pages. Also French-English vocabulary list, exercises, more. 352pp. 5⅜ x 8½. 26443-2 Pa. $9.95

CHICAGO AT THE TURN OF THE CENTURY IN PHOTOGRAPHS: 122 Historic Views from the Collections of the Chicago Historical Society, Larry A. Viskochil. Rare large-format prints offer detailed views of City Hall, State Street, the Loop, Hull House, Union Station, many other landmarks, circa 1904-1913. Introduction. Captions. Maps. 144pp. 9⅜ x 12¼. 24656-6 Pa. $12.95

OLD BROOKLYN IN EARLY PHOTOGRAPHS, 1865-1929, William Lee Younger. Luna Park, Gravesend race track, construction of Grand Army Plaza, moving of Hotel Brighton, etc. 157 previously unpublished photographs. 165pp. 8⅞ x 11¾. 23587-4 Pa. $13.95

THE MYTHS OF THE NORTH AMERICAN INDIANS, Lewis Spence. Rich anthology of the myths and legends of the Algonquins, Iroquois, Pawnees and Sioux, prefaced by an extensive historical and ethnological commentary. 36 illustrations. 480pp. 5⅜ x 8½. 25967-6 Pa. $10.95

AN ENCYCLOPEDIA OF BATTLES: Accounts of Over 1,560 Battles from 1479 B.C. to the Present, David Eggenberger. Essential details of every major battle in recorded history from the first battle of Megiddo in 1479 B.C. to Grenada in 1984. List of Battle Maps. New Appendix covering the years 1967-1984. Index. 99 illustrations. 544pp. 6½ x 9¼. 24913-1 Pa. $16.95

SAILING ALONE AROUND THE WORLD, Captain Joshua Slocum. First man to sail around the world, alone, in small boat. One of great feats of seamanship told in delightful manner. 67 illustrations. 294pp. 5⅜ x 8½. 20326-3 Pa. $6.95

ANARCHISM AND OTHER ESSAYS, Emma Goldman. Powerful, penetrating, prophetic essays on direct action, role of minorities, prison reform, puritan hypocrisy, violence, etc. 271pp. 5⅜ x 8½. 22484-8 Pa. $7.95

MYTHS OF THE HINDUS AND BUDDHISTS, Ananda K. Coomaraswamy and Sister Nivedita. Great stories of the epics; deeds of Krishna, Shiva, taken from puranas, Vedas, folk tales; etc. 32 illustrations. 400pp. 5⅜ x 8½. 21759-0 Pa. $12.95

BEYOND PSYCHOLOGY, Otto Rank. Fear of death, desire of immortality, nature of sexuality, social organization, creativity, according to Rankian system. 291pp. 5⅜ x 8½. 20485-5 Pa. $8.95

A THEOLOGICO-POLITICAL TREATISE, Benedict Spinoza. Also contains unfinished Political Treatise. Great classic on religious liberty, theory of government on common consent. R. Elwes translation. Total of 421pp. 5⅜ x 8½. 20249-6 Pa. $9.95

PHOTOGRAPHIC SKETCHBOOK OF THE CIVIL WAR, Alexander Gardner. 100 photos taken on field during the Civil War. Famous shots of Manassas Harper's Ferry, Lincoln, Richmond, slave pens, etc. 244pp. 10⅝ x 8¼. 22731-6 Pa. $10.95

FIVE ACRES AND INDEPENDENCE, Maurice G. Kains. Great back-to-the-land classic explains basics of self-sufficient farming. The one book to get. 95 illustrations. 397pp. 5⅜ x 8½. 20974-1 Pa. $7.95

SONGS OF EASTERN BIRDS, Dr. Donald J. Borror. Songs and calls of 60 species most common to eastern U.S.: warblers, woodpeckers, flycatchers, thrushes, larks, many more in high-quality recording. Cassette and manual 99912-2 $9.95

A MODERN HERBAL, Margaret Grieve. Much the fullest, most exact, most useful compilation of herbal material. Gigantic alphabetical encyclopedia, from aconite to zedoary, gives botanical information, medical properties, folklore, economic uses, much else. Indispensable to serious reader. 161 illustrations. 888pp. 6½ x 9¼. 2-vol. set. (USO)
Vol. I: 22798-7 Pa. $9.95
Vol. II: 22799-5 Pa. $9.95

HIDDEN TREASURE MAZE BOOK, Dave Phillips. Solve 34 challenging mazes accompanied by heroic tales of adventure. Evil dragons, people-eating plants, blood-thirsty giants, many more dangerous adversaries lurk at every twist and turn. 34 mazes, stories, solutions. 48pp. 8¼ x 11. 24566-7 Pa. $2.95

LETTERS OF W. A. MOZART, Wolfgang A. Mozart. Remarkable letters show bawdy wit, humor, imagination, musical insights, contemporary musical world; includes some letters from Leopold Mozart. 276pp. 5⅜ x 8½. 22859-2 Pa. $7.95

BASIC PRINCIPLES OF CLASSICAL BALLET, Agrippina Vaganova. Great Russian theoretician, teacher explains methods for teaching classical ballet. 118 illustrations. 175pp. 5⅜ x 8½. 22036-2 Pa. $5.95

THE JUMPING FROG, Mark Twain. Revenge edition. The original story of The Celebrated Jumping Frog of Calaveras County, a hapless French translation, and Twain's hilarious "retranslation" from the French. 12 illustrations. 66pp. 5⅜ x 8½. 22686-7 Pa. $3.95

BEST REMEMBERED POEMS, Martin Gardner (ed.). The 126 poems in this superb collection of 19th- and 20th-century British and American verse range from Shelley's "To a Skylark" to the impassioned "Renascence" of Edna St. Vincent Millay and to Edward Lear's whimsical "The Owl and the Pussycat." 224pp. 5⅜ x 8½. 27165-X Pa. $5.95

COMPLETE SONNETS, William Shakespeare. Over 150 exquisite poems deal with love, friendship, the tyranny of time, beauty's evanescence, death and other themes in language of remarkable power, precision and beauty. Glossary of archaic terms. 80pp. 5³⁄₁₆ x 8¼. 26686-9 Pa. $1.00

BODIES IN A BOOKSHOP, R. T. Campbell. Challenging mystery of blackmail and murder with ingenious plot and superbly drawn characters. In the best tradition of British suspense fiction. 192pp. 5⅜ x 8½. 24720-1 Pa. $6.95